T0136423

Cryptography and Network Security

RIVER PUBLISHERS SERIES IN SECURITY AND DIGITAL FORENSICS

Series Editors:

ANAND R. PRASAD
Deloitte Tohmatsu Cyber LLC in
Japan

R. CHANDRAMOULI
Stevens Institute of Technology
USA

ABDERRAHIM BENSLIMANE
University of Avignon
France

PETER LANGENDÖRFER
Brandenburg University of Technology,
Leibniz Institute for High Performance
Microelectronics (IHP)
Germany

The "River Publishers Series in Security and Digital Forensics" is a series of comprehensive academic and professional books which focus on the theory and applications of Cyber Security, including Data Security, Mobile and Network Security, Cryptography and Digital Forensics. Topics in Prevention and Threat Management are also included in the scope of the book series, as are general business Standards in this domain.

Books published in the series include research monographs, edited volumes, handbooks and textbooks. The books provide professionals, researchers, educators, and advanced students in the field with an invaluable insight into the latest research and developments.

Topics covered in the series include, but are by no means restricted to the following:

- Cyber Security
- Digital Forensics
- Cryptography
- Blockchain
- IoT Security
- Network Security
- Mobile Security
- Data and App Security
- Threat Management
- Standardization
- Privacy
- Software Security
- Hardware Security

For a list of other books in this series, visit www.riverpublishers.com

Cryptography and Network Security

Marcelo Sampaio de Alencar

Institute of Advanced Studies in Communications
University Center Senai Cimatec, Salvador

Routledge
Taylor & Francis Group

LONDON AND NEW YORK

Published 2022 by River Publishers
River Publishers
Alsbjergvej 10, 9260 Gistrup, Denmark
www.riverpublishers.com

Distributed exclusively by Routledge
4 Park Square, Milton Park, Abingdon, Oxon OX14 4RN
605 Third Avenue, New York, NY 10017, USA

Cryptography and Network Security / by Marcelo Sampaio de Alencar.

Routledge is an imprint of the Taylor & Francis Group, an informa business

ISBN 978-87-7022-407-9 (print)
ISBN 978-10-0079-293-5 (online)
ISBN 978-1-003-33776-8 (ebook master)

While every effort is made to provide dependable information, the publisher, authors, and editors cannot be held responsible for any errors or omissions.

I dedicate this book to my family.

Contents

Preface

"The enemy knows the system."

Claude Shannon

"Cryptography and network security" is a hot topic in the educational market. It evolved from the amalgamation of different areas of mathematics, logic, computer networks, probability, and stochastic processes, and includes the beautiful development of Communication Theory by Claude Shannon, in 1948.

The area of cryptography and network security is fundamental to several branches of knowledge, including engineering, computer science, mathematics, physics, sciences, economics, social sciences, and social communication. It is part of the syllabus for most courses in computer science, mathematics, and engineering.

For electrical engineering courses, it follows some disciplines, including communication systems, transmission techniques, error control coding, estimation, and digital signal processing. This book is self-contained, and it is a reference and an introduction for graduate students who did not take information theory before. It could also be used as an undergraduate textbook.

It is addressed to a large audience in electrical and computer engineering, or mathematics and applied physics. The book targets graduate students and engineers in these areas, who may not have taken basic courses in specific topics, and who will find a concise way to obtain the knowledge they need to succeed in advanced courses.

According to a study published by the Institute of Electrical and Electronics Engineers (IEEE), the companies, enterprises, and industry are in need of professionals with a solid background on mathematics and sciences, instead of the specialized professional of the previous century. The employment market in this area is in demand of information technology professionals and engineers who could afford to change and learn as the market changes. The market needs professionals who can model and design.

In this regard, few books have been published covering the subjects needed to understand the very fundamental concepts of cryptography and network security. Most books that deal with the subject are usually destined to very specific audiences.

The more mathematically oriented books are seldom used by people with engineering, economics, or statistical background because the authors are more interested in theorems and related conditions than in fundamental concepts and applications. The books written for engineers and software professionals usually lack the required rigor, or skip some important points in favor of simplicity and conciseness.

The idea is to present a seamless connection between the more abstract advanced Information Theory and the fundamental concepts of coding theory and computer networks and protocols, filling in the gaps from previous books and leading to an interesting, robust, and, hopefully, self-contained exposition of cryptography and network security.

The book begins with the historical evolution of networks and security in Chapter 1. The main definitions to understand security systems, Internet attacks, and cryptography are presented in Chapter 2. Chapter 3 deals with the basic concepts of information theory, and how to measure information. The usual types of source codes are presented in Chapter 4. Information transmission, joint information, conditional entropy, mutual information, and channel capacity are the subject of Chapter 5. Computer networks are discussed in Chapter 6.

Chapter 7 presents the main protocols and network architectures. The important TCP/IP protocol is the subject of Chapter 8. Network security, a topic intrinsically connected to computer networks and the Internet, is presented in Chapter 9, along with the basic hacker attacks, alternatives to prevent attacks, data protection, and secure protocols.

The information theoretical aspects of cryptography are presented in Chapter 10. Chapter 11 introduces the hash function. Some interesting cases of cyberattacks are discussed in Chapter 12.

Appendix A includes a review of probability theory, with illustrations and graphics that help the reader understand the theory. Appendix B presents the basics of cryptoalgorithms and cryptosystems. Appendix C includes a complete glossary of information security terms. A short biography of the author is presented before the bibliography. The book also includes a remissive index to help the readers find the location of the adequate information.

Marcelo S. Alencar

Acknowledgements

The publication of this book is the result of the experience acquired by the author throughout the years, combined with the knowledge received from professors and conveyed by book authors.

This book evolved during the many years of the author's work at the Federal University of Campina Grande (UFCG), the Federal University of ParaÃba (UFPB), the University for the Development of the State of Santa Catarina (UDESC), the Federal University of Bahia (UFBA), the Senai Cimatec University Center, Salvador, and the Institute of Advanced Studies in Communications (Iecom).

It has been a privilege to cooperate with several companies, firms, and institutions, for many years, including Embratel, Atecel, Correios do Brasil, Telern, Chesf, and Telpa. This cooperation also involved the companies Telpe, TIM, Siemens, Oi, Telebrás, Alpargatas S/A, Licks Attorneys, and the National Telecommunications Agency (Anatel).

The author thanks the authoritative translation of some of the chapters by Thiago Tavares de Alencar, and the complete revision of the text performed by Junko Nakajima. Any mistakes, found perchance in the text, are exclusive responsibility of the author.

The understanding and affection of Silvana, Thiago, Raissa, Raphael, Janaina, Marcella, Vicente, and Cora, who hopefully forgave the long periods of absence on account of the academic work, conferences, and sabbatical periods, allowed the author to develop this book, based on articles published in journals, magazines, and conferences.

Marcelo S. Alencar

List of Figures

List of Tables

1

Introduction

> "A cryptographic system should be secure even if everything about the system, except the key, is public knowledge."

> Auguste Kerckhoffs

The concept of information, in its journalistic or social aspect, was established initially by Charles Sanders Peirce (1839–1914), an American philosopher, linguist, and mathematician, who created the field of Semiotics, the study of symbols. Peirce also defined the amount of information in an object, such as, for example, a scene, from the number of words, or characters, used to its description.

Later, in the 1960s, Andrei Kolmogorov (1903–1987), a Russian mathematician and engineer, created the equivalent concept of complexity, related to the number of lines of programming, in a Turing machine to fully describe a particular event or object. The area evolved to become the computational complexity theory, that focuses on the classification of computational problems, according to their resource usage, and relating these classes to each other.

The Mathematical Theory of Information was developed by Claude E. Shannon (1916–2001), an American mathematician, electronics engineer, and cryptographer, from an article published in 1948 (Shannon, 1948a), in which he establishes the main definitions, premises, and rules of the theory and also deduces the main theorems.

Information security is based on two main axes of the theory: the calculation of the capacity of the channels, used to determine the maximum transmission rate of information in communication networks, and the computation of the uncertainty related to the generation of cryptographic keys.

An important fact to consider is that the virtual economy is real (Alencar, 2015). Business is conducted over the Internet with increasing intensity each year. The evildoers are on the network; they are the hackers (those who aim to penetrate secure systems just for pleasure) and the crackers (criminals who steal passwords and money from user accounts), which disrupt Internet users' lives. The inconveniences are on the network and give rise to spam (those unsolicited messages, usually advertising) and disclose hoaxes (the untrue messages that scare the novice Internet users).

For centuries, leaders, traders, and the military, among others, have been concerned with the preservation of sensitive information, i.e., information that should only be seen by specific recipients, and algorithms have been developed to hide this type of information from curious eyes.

A famous puzzle, published by the English mathematician and author Henry Dudeney (1857–1930) in The Strand Magazine, illustrates how information can be hidden in an apparently innocent text (Dudeney, 1924).

In the presentation of the puzzle, Dudeney says that a kidnapper asked for a certain amount in cash, sending a coded text, which appeared as if it were a sum of two installments. The goal was to find the requested amount.

$$SEND$$
$$MORE$$
$$MONEY$$

The solution of the cryptogram can be obtained by assembling a system of equations, in which the letters, the variables of the problem, are considered.

$$D + E = Y$$
$$N + R = E + V1$$
$$E + O = N + V2$$
$$S + M = O + V3$$
$$M = V4$$

in which V_i represents the decimal excess (goes one) of each partial sum. Solving the system of equations, one obtains

$$9567$$
$$1085$$
$$\cdots$$
$$10,652$$

A curious detail about The Strand Magazine is that Sir Arthur Conan Doyle (1959–1930), the British writer, physician, politician, poet, and creator of the most famous detective in police literature, Sherlock Holmes, contributed for decades to the magazine, with novels, short stories, poems, and articles.

1.1 Information on the Network

Computer networks store much of the world's information, and that data needs to be searched for and processed to be useful, while the information must also be kept in security to guarantee its integrity.

Big data is a term associated with data sets that are too large or complex to be processed by usual data processing programs or applications, in a timely manner, at a reasonable cost (Alencar, 2017c).

These huge databases are often complicated to capture, store, share, search for information, transfer data, visualize information, update, maintain privacy, analyze, check, and also inquire about a certain information.

The term big data is also used to designate the procedure of predictive analysis, behavioral analysis, or other analytical methods of extracting value from data sets.

Big data examples include the Internet itself, financial data sets, data from urban computing or business computing, data obtained from sensor networks in smart cities, and data from the Internet of things, to name a few.

Usually, these large data sets are examined to discover hidden patterns, unknown correlations, market trends, consumer preferences, and other information that can be useful to assist organizations in making business decisions, in an intelligent or well-informed manner.

For example, Google Analytics is used to analyze the use of web pages because it produces statistical information about the use of the websites by searching for stored cookies. From this extraction, or data mining, several characteristics of the users are revealed.

Currently, it is not uncommon for a given network user to receive information about a specific city, including how to book hotels, how to purchase tickets, and what are the best tours, as a result of a simple visit to the city web page.

At the end of 2016, the International Telecommunication Union (ITU) already accounted for nearly four billion people connected to the Internet. Research in such a large universe, with such diversity of information, certainly requires big data techniques.

1.2 The Internet of Things

There is an expectation that all the usual objects will be connected to the Internet in a few years and that they will be able to operate in a sensory and intelligent way, an anthropomorphism widely used in the area, but that may not be far from reality (Alencar, 2017b).

The Internet of Things (IoT) consists of the fusion of the real world, or physical, with the virtual, or computational, which can allow perennial communication and productive interaction between human beings, animals, and objects.

There are many possibilities to add devices to the network, which may include biological function monitoring equipment, sensors and sensor networks, automobiles, electronic keys, appliances, and clothing, among other things.

The technological revolution caused by the inclusion of objects in the network is considered one of the last stages of the development of computing, as the possibility of leading the world to singularity, that is, to the point where, according to some scientists, the network will become self-conscious.

British researcher Kevin Ashton (1968-), from the Massachusetts Institute of Technology (MIT), first used, in 1999, the term Internet of Things to refer to the use of technologies that interconnect devices and objects to facilitate and organize people's lives. Some examples of the possible uses of the new technology are listed in the following.

Sensors connected to the network can indicate exactly where, when, and how a product can be used, to assist in design and marketing processes. Real-time data collection can be faster, more accurate, and less costly than market research.

Problems associated with manufacturing, that is, the condition, definitions, and use of production equipment, can be minimized by monitoring the production process, so that errors are identified, losses are reduced, and corrective actions are taken to increase uptime and business efficiency.

The network theory, a recent discipline, was born from the need to understand the nature of large networks and the search for characteristics common to the interconnection of intelligent entities, or not, to form structures with new properties, such as the World Wide Web. This theory can indicate how close the Internet of Things would be to the fictional Skynet.

The Internet of Things (IoT) has the potential to connect tens of billions of computers, sensors, devices, and networks worldwide, while creating new challenges for data security circulating in this network (Alencar, 2017a).

The expression Internet of Things was born in 1999 from the idea of electronically labeling the products of a company to facilitate the logistics of the production chain, with the use of radio frequency identifiers (in English, RFID), a really new subject, in era.

The monitoring of the condition and the use of connected components can indicate, in addition to prices, when consumers will demand replacement parts, ensuring that there are suitable products, and available, at the supplier.

Component wear information can help reduce maintenance and operating costs as well as identify potential equipment failures before production is stopped.

Monitoring the condition of machines and equipment, including configurations and conditions of use, can improve product engineering in terms of material choice and design.

The installation of sensor networks in large delivery containers allows the receipt of real-time data on the location of the packages, the frequency of handling, and the condition of the product. The inventory management system can use this data to improve logistics, increase efficiency, accelerate delivery times, and improve customer service.

Transport companies can offer services based on Internet of Things applications to contribute to the creation of smart cities, another anthropomorphism that has been successful, with the administration of parking spaces and the management of bus schedules, for example.

Wireless sensor networks can be used to monitor speed, fuel economy, mileage, number of stops, and engine wear in fleets. Thus, repairs can be scheduled, avoiding interruptions in logistics, and behaviors that increase fuel consumption can be identified. In addition, vehicle maintenance and proper driving can decrease CO_2 emissions and increase vehicle life expectancy.

Wireless sensor networks can monitor air temperature, soil conditions, wind speed, humidity, and solar radiation, for example, and use information from the network, such as the likelihood of rain, to determine the ideal time to irrigate the plants. Precision agriculture is beginning to benefit from network engineering.

Doctors and hospitals can receive and organize data from medical devices connected to the network, which can be installed in hospitals or in patients' homes. With real-time information, it is possible to improve care through more effective diagnoses and treatments.

The Internet of Things has the potential to connect tens of billions of computers, sensors, devices, and networks worldwide. With the integration of people and machines, the virtual potential of the Web also becomes physical, with the possibility of network processes, devices, and robots acting on real machines, vehicles, and equipment.

1.3 Network Security

The Internet has changed the way companies do business, as the Internet Protocol (IP), developed in the 1970s by Robert E. Kahn (1938-) and Vint Cerf (1943-), made data transmission efficient, cheap, and flexible. However, a poorly managed or poorly configured network is vulnerable to pernicious attacks, which can cause loss of information, network resources, or money (Alencar, 2018c).

Thus, it is necessary to guarantee the company's business, which is the most precious asset for its operation, or the user's well-being, which allows tranquility, protecting information. For this, it is necessary to take into account the fundamentals of information security in the design of the local network.

First, it is necessary to guarantee the integrity of the data, that is, to make sure that the information is not modified, between the source and the destination, during the communication. This can even be done with the support of a certification body.

Confidentiality must be maintained; in other words, the information must be restricted to only those parts authorized by the management of the network. This can lead to the encryption of the data to be transmitted, as an alternative.

The guarantee that the information comes from a legitimate source is known as authenticity, a service that can also be provided by an external entity. And the provision of the service must be made only to a legitimate user, guaranteeing the availability of this service.

The set of measures to prevent the sender from denying the transmitted message or the recipient from denying the received message, which can cause a problem between the parties, is known as non-repudiation.

It is important to ensure accountability, that is, the characteristic of the system that allows the provision of audit trails for all transactions, and access control, to ensure that access is allowed only to authorized users, through identification and authorization.

With these security practices, the local network will be protected and the Internet will have a weak link for invasion, data theft, or insertion of malicious programs.

1.4 Network Encryption

According to the Encyclopedia Britannica, cryptography is the practice of the enciphering and deciphering of messages in secret code in order to render them unintelligible to all but the intended receiver. Cryptography may also refer to the art of cryptanalysis, by which cryptographic codes are broken (Hosch et al., 2021).

In other words, cryptography is the process of converting between readable text, called plaintext, and an unreadable form, called ciphertext (IBM, 2021).

The cryptographic process involves three distinct phases, as described in the following:

- The sender, who wants to send a secret information to a receiver, converts the plaintext message to a ciphertext. This phase is called encryption or encipherment.
- The ciphertext is transmitted to the receiver, who has agreed on a common cipher, previously, with the receiver.
- The receiver converts the ciphertext message back to its plaintext form, using the cipher. This part of the process is called decryption or decipherment.

Computer networks are constantly under attack by attackers, often known as hackers or crackers. Network security professionals are concerned, therefore, with defining some objectives to guarantee the integrity of the systems (Alencar, 2018b).

It is important to control attacks that threaten the security goals set by the company or institution to provide security services and their interrelationships and to ensure mechanisms to provide security services and to introduce encryption techniques.

The original message, to be transmitted, is identified as clear text and will be converted to a sequence with a random appearance so as not to be easily recognized, called cipher text.

The encryption process uses an encryption algorithm that depends on a password, previously chosen, preferably without connection to the original text. With each new password used, the algorithm produces a new output.

It is considered that a cryptanalyst, or hacker, can observe the encrypted message and try to decode it. Therefore, the algorithm must resist attacks. At reception, an inverse transformation is performed on the ciphertext to produce the clear text.

For that, the receiver must have the key that inverts the transformation, and this implies that the generation, transmission, and maintenance of the key are important parts of the encryption process.

There are two ways to encrypt text. There is encryption with a simple key and encryption that uses a public key and another private key. The unique key, previously agreed between the parties, is used to encrypt information and also to recover it.

Encryption with a public key and another private key ensures greater security for the transmission of information because there is no need for prior agreement between the parties. The public key is released on the network and is used to encrypt the information. The private key is kept secret and is used to decrypt the message.

The security of a cryptographic system, then, depends on some conditions: the encryption algorithm must be robust, to make it difficult to decrypt the message, based only on the ciphertext, the key must be kept secret, and there must be a secure channel for its transmission; but, interestingly, it is not necessary to keep the algorithm secret.

One of the first women to develop a cryptographic system was the Austrian actress, naturalized American, Hedwig Eva Maria Kiesler (1914–2000), known by her stage name Hedy Lamarr. She and composer George Antheil developed a communications system based on a technique that would known as spectral spreading, which was used in cellular mobile communications standards.

1.5 Electronic Voting

Brazil was one of the first countries to conduct electronic voting. The Superior Electoral Court adopted electronic voting in the 1996 elections. The 2000 elections already had electronic ballot boxes throughout the country, with the Linux operating system installed as of 2008 (Alencar, 2018a).

Engineers and researchers linked to the General Command of Aerospace Technology (CTA) and the National Institute for Space Research (INPE) participated in the large-scale computerized election project in the country.

The electronic ballot box, which has already automated 100% of the elections in Brazil, was developed by the company OMNITECH Serviços em Tecnologia e Marketing, between 1995 and 1996, and improved in 1997, reaching the model that became the Brazilian standard.

The main components of the Brazilian electronic ballot box include the memory, which has two cards with identical data, to minimize the risk of failures, in which the operating system, the application programs, the data about the candidates, and the votes are recorded, with the use of security and redundancy mechanisms to make fraud more difficult.

There is also a flash drive, to record the result at the end of the vote, a printing module, used before the start and at the end of the vote to print the ballot box, and a polling station terminal, which has a numeric keypad with a biometric reader of the voter's fingerprint, used for the polling authority to authorize the voter to vote.

The voter terminal, by which the voter chooses his candidates, consists of a liquid crystal display and a numeric keypad, with keys for correction, confirmation, and blank voting.

Six months before the election, the Superior Electoral Court (TSE) presents the source codes to political parties, the Brazilian Bar Association (OAB), and the Public Ministry (MP). In this event, the source codes, specifications, and program documentation are presented.

During the code opening period, systems can be inspected using code analysis tools. Two of them are made available by the TSE: Understanding C and Source Navigator.

The meeting is documented and serves to improve the process, the program, or the equipment because the representatives can request improvements, ask questions, or talk to the TSE technical team.

Although no electoral process in the world is inviolable or unquestionable, whatever its nature, electronic voting minimizes the occurrence of fraud because there is a password that encrypts the results, and few people are able to understand the codes used. Unlike paper voting, it can be defrauded by anyone.

The curious thing is that, recently, some unscrupulous politicians have spread rumors that the polls would have problems or would be insecure, suggesting as a solution the convenient impression of the vote.

However, in addition to the impression of voting being a strange practice to the electoral process, never having been carried out in the country, not even when the vote was on paper, this procedure appears to be just a way of guaranteeing the politician who buys votes a receipt of the purchased vote.

1.6 Security with Biometrics

The use of fingerprints to identify people was an ancient practice in many places, such as Mesopotamia, Turkestan, India, Japan, and China. In antiquity, it aimed to authenticate documents and establish civil and commercial agreements, since only the noble and wealthy knew how to read and write (Alencar, 2017e).

The first uses of digital biometrics for identification were associated with prehistory, more specifically with Babylon. In China, some artisans printed their fingerprints on vases, as a way to give identity to their works.

The fingerprint identification system, with forensic applications, was devised by Francis Galton (1822–1911), a British meteorologist, mathematician, and statistician, who also created the concept of statistical correlation. Galton was a cousin of Charles Darwin (1809–1882) and, based on his research, created the concept of eugenics, that is, the improvement of a certain species through artificial selection.

Eugenics was used to sterilize thousands of women, who had children with any type of disability, in the United States of America, and was used by Adolf Hitler to justify the elimination of Jews in Nazi Germany.

Supporters of capitalism have also abused the concept of eugenics, to promote the system of economic liberality, known as laissez-faire, in which it is argued that the market must operate freely, without interference from the state, which would have only sufficient regulations to protect property rights.

For these neo-liberals, only the best would be able to establish themselves, to the detriment of the poorest, who would be in this condition because of some intrinsic deficiency. The wealthier, certainly, would have a preference in some artificial selection process to improve the economic genome, and so they preached.

Back in 1903, in New York City, the fingerprint began to be collected in a systematic way, as it is a characteristic of each individual, being different even between identical twins, to set up a database with the objective of identifying criminals.

The fingerprint, technically known as a typewriter or dermatoglyph, is the design formed by the papillae, which are the elevations of the skin, and the interpapillary grooves, present in the pulps of the fingers. They are generally perennial, immutable, and have different shapes, for different people.

Today, digital printing is used as a biometric identification at ATM, to release money or provide services, to open doors, to guarantee the integrity of electronic voting, among other applications.

The curious thing is that Francis Galton, who was a sociologist, as well as a polymath, anthropologist, statistician, geographer, meteorologist, and psychologist, having published more than 340 books and scientific articles, believed that the fingerprint could identify the race of a certain individual. It was certainly used to identify and select Jews who would go to concentration camps and to gas chambers.

Some current technologies, which could be used in the security systems of shopping centers and banks, include biometric recognition of the iris, the most visible and colorful part of the eye, face, and behavior (Alencar, 2017d).

Biometrics is the statistical study of the physical or behavioral characteristics of living beings. But the term has been associated with the measurement of physical or behavioral characteristics of individuals, as a way of identifying them in a unique way.

It has been used, for example, in criminal identification and in controlling access to environments or services. Biometric systems can use features from different parts of the human body, including the eyes, the palm of the hand, the fingerprints, the retina, the part on which the image is projected on the eye, or the iris.

Some of the potential biometric reading systems include blood vessels, veins, and arteries, which are of average reliability, difficult to defraud, but expensive. Voice recognition is also feasible, but it is less reliable, as it presents problems with noise in the environment and also with changes in the voice, in addition to the high processing time for recording and reading, but it has low cost.

The writing style, a type of behavioral biometry, can also be used for identification, but it can be falsified. The odors and salinity of the human body also have potential for use as well as thermal images of the face or the other part of the body, and DNA analysis, a technique with high reliability but, currently, of high cost and long processing time.

The geometry of the hand is less reliable mainly due to tattoos and the use of jewelry, in addition to the work of fitting the hand in the correct position on the sensor, but it has medium cost. Face recognition is less reliable, but it is fast and inexpensive. Of course, it can be distorted with surgery.

Signature recognition has been used by banks for a long time because it is very reliable. But, some signatures change over time. However, characteristics such as pressure, particular movements, are unique to each person, which makes counterfeiting difficult. The method is quick, practical, accessible to everyone who can read, and has a medium cost.

1.7 How to Understand Recognition

Biometric solutions, generally developed at universities and incorporated by companies into their systems, recognize individual attributes, such as the shape of the face, the timbre of the voice, or any physical and behavioral characteristics of human beings.

This market is dynamic and is expected to generate more than US\$ 30 billion in 2021 worldwide, according to the American consulting company ABI Research.

The banking sector always needs to improve its customer identification processes and already has 90 thousand ATMs equipped with biometric sensors. Two-thirds of these ATMs have multispectral sensors to identify deeper layers of tissue, including the visualization of blood vessels, which makes it difficult to use fake fingers, latex, or plastic, for example.

The creation of a civil identification (CI) card should encourage the use of biometrics with the incorporation of the main public data and individual biomedical records in an integrated circuit, which will also be incorporated into the Electoral Justice database.

The registration, analysis, and validation of identity in different biometric processes are similar. In general, the same structure is used, regardless of the body part used. This occurs so that the systems form accessible and quickly analyzed databases, which facilitates the verification of the individual's identity.

The process begins with the capture, which represents the phase of recording the data to be used to prove identity. In general, the procedure requires that the indicator be positioned on the optical reader or that a standard word or sentence be repeated. This process can be repeated until the registration is reliable.

The second phase is extraction, in which the collected data is translated into information that can be identified by the biometric system. There is some variation in the procedure, as each system has a particular method of translation, which varies in reliability and precision, to transform the image into a file that can be read by the machine.

After translating the information into the computer's language, the system creates a standard for registration, based on characteristics recognizable by the biometric system. This defines the initial standard, which is converted to the final format, for storage. With this, image data is saved in a simpler file, easier to be analyzed by the system, which reduces the analysis time and the necessary memory.

With the registration and creation of the standard, the comparison is performed to determine the efficiency of registration of the fundamental information. If there are identification failures, or false positive identifications, due to the low quality of the information collected, the entire process is redone in order to obtain consistent and reliable results.

The perverse side of biometrics is that it aims at the complete automation of banks, replacing the recognition made by human tellers, whose positions will soon be extinguished. It is worth mentioning that banks have already passed on part of the tasks that were performed by their employees to customers, who need to understand the banking procedures to carry out their transactions from their personal computers or smartphones.

However, the productivity gains obtained were not passed on to customers, nor did they result in a decrease in bank fees, which are generally high and continue to be paid, even if the customer never needs to go to the bank, and even if the bank does not even have branches.

1.8 Blockchain and Cryptocurrency

A blockchain is a shared, immutable ledger that records transactions and tracks assets in a business network. An asset can be tangible, such as a house or a car, or intangible, such as an intellectual property, a patent, or a copyright. Most valuable assets can be tracked and traded on a blockchain network (IBM, 2021).

1.8.1 Enterprises Can Profit from Blockchain

The blockchain concept is important because most businesses run on information and blockchain can deliver the information that is stored on an immutable ledger, accessible only by permission of network members, in an immediate, shared, and completely transparent manner. A blockchain network can track orders, payments, accounts, and production.

A blockchain network can be public, private, permissioned, or built by a consortium, and the principal elements of a blockchain are the distributed ledger technology, the immutable records, and the smart contracts.

All network participants have access to the distributed ledger and its immutable record of transactions. Transactions are recorded only once in this ledger, eliminating the duplication of effort and data that is typical of common business networks.

No participant can change or adulterate a transaction after it has been recorded to the shared ledger. If a transaction record includes an error, a new transaction must be added to reverse the error, and both transactions are then visible by the users.

A set of rules, called a smart contract, which is stored on the blockchain and executed automatically, is used to speed up transactions. A smart contract can define conditions for corporate bond transfers, including terms for travel insurance to be paid, for example.

Anyone can join and participate in a public blockchain, such as Bitcoin. The usual problems are related to the computational power required, lack of privacy for transactions, and weak security, which are important considerations when an enterprise uses a blockchain.

A private blockchain network is a decentralized peer-to-peer network, similar to a public blockchain network. However, one organization governs the network, controlling who is allowed to participate, execute a consensus protocol, and maintain the shared ledger. This can increase trust and confidence between participants mainly because a private blockchain run behind a corporate firewall or be hosted on premises.

Businesses who establish a private blockchain are prone to assemble a permissioned blockchain network. This places restrictions on who is allowed to participate in the network and can restrict the allowed transactions because the participants need to obtain an invitation or permission to join. Public blockchain networks can also be permissioned.

Multiple organizations can share the responsibilities of maintaining a blockchain. These pre-selected organizations determine who may submit transactions or access the data. A consortium blockchain is adequate for business when all participants need to be permissioned and have a shared responsibility for the blockchain (IBM, 2021).

1.8.2 The Cryptocurrency Frenzy

In 2009, a character called Satoshi Nakamoto, whose real identity continues to be unknown, developed the reference for the implementation of an encrypted virtual coin, known as Bitcoin (BTC), that uses the blockchain, a process of data sharing among several parts, which is apparently inviolable (Alencar, 2021).

Bitcoins are negotiated and stored in a decentralized network of computers, which is not under the control of any government or company. An attractive aspect for investors from countries with a history of savings, accounts, and assets sequestration, but also an obscure port for tax evaders, drug dealers and corrupt people in general.

For those who like history, and value their assets, it is worth knowing that the virtual crypto coin Bitcoin reached the stratospheric value of US$ 63 thousand in April 2021. That is correct, a single Bitcoin, a binary encrypted code with no real currency backing, was worth 200–250 salaries of a specialized technician. For the record, the alleged creator of the currency disappeared in 2010.

Cryptocurrency is a type of virtual coin that uses encryption to guarantee safer financial transactions on the Internet. There are several types of crypto coins. Bitcoin is the best known, a digital encrypted coin which allows financial transactions free of institutions, but monitored by users of the network and encoded in a (blockchain) database.

Bitcoin does not derive from physical currency; it is not backed financially and is not recognized by the Securities and Exchange Commission (CVM) in Brazil as an asset. Despite being introduced in 2009, the de facto asset Bitcoin (BTC) only aroused interest in the media in 2012. Strictly speaking, the encrypted financial assets, or crypto assets, should not be classified as currency since they do not meet the requirements to receive this denomination, considering the current monetary theory.

Currency is typically defined by three fundamental attributes: functioning as a means of trade, being a unit of counting, and acting as a value reserve. Bitcoin meets the first criteria because a growing number of merchants, especially in online markets, are willing to accept it as a means of payment. However, the commercial worldwide use of Bitcoin remains reduced, indicating that few people are using it as a means of trade (Camacho and da Silva, 2018).

Blockchain is a decentralized digital database, which registers financial transactions that are stored in computers throughout the world. The database registers the sending and the receiving of the values of encrypted format digital coins, and the parties need to authorize the access among themselves.

Bitcoin originated from a process developed by Satoshi Nakamoto, who described the development of a peer-to-peer (P2P) electronic money system. The algorithm proposed by Nakamoto creates new Bitcoins and awards them to computer users that resolve specific mathematical problems. These problems get more complex and less frequent with time, in function of the costs related to mining, which is the process of generating new coins.

Miners are individuals or companies that are involved in the activity to win new blocks of Bitcoins. The payments are registered in a public ledger book, and the transactions occur peer-to-peer, without a central repository or unique administrator.

This decentralized record monitors the property and the subsequent transfers of each Bitcoin, after they are extracted, or mined, by their original proprietors. The ledger is organized as a blockchain, that contains records of validated transactions to track the property of each Bitcoin.

Each transaction record contains a public key for the recipient. In a Bitcoin transaction, the proprietor validates his property using a private key and sends an encrypted instruction with his key. The system records the transaction instruction, which contains the public key of the new owner, which is the recipient in a new block. This system of decentralized authentication also serves for dealing with falsifications and double accounting problems.

The P2P network attributes parity of rights to all the members. The decentralization is maintained combining Proof of Work (PoW) with other cryptographic techniques. PoW is a hash mathematical function, an algorithm that maps variable length data into fixed length data, that has a large number of possible results for each input.

The hash in each block is connected to the next block. The hash function can be represented in the following manner:

$$h(n) = f[h(n-1), \phi, K], \tag{1.1}$$

in which $h(n)$ represents the hash of the current block, $h(n-1)$ is the hash of the previous block, ϕ is the difficulty level, and K is the random specific key for the current block.

This indicates that each subsequent block is connected to the previous one. If there is a dishonest miner that decides to generate an invalid block, the other members of the network will not confirm the operation because the hash of the previous block will be validated.

Even if a user tries to change the hash of the previous block, he would have to do it for the block before it and successively until the very first block created by Nakamoto. This would require an enormous amount of computational work, which would exceed the current capacity of the network.

So, to protect the integrity of the ledger book, the system protects each block with a unique hash. The hash is generated based on the system information and the owner's key and needs to meet the criteria given by the hashrate defined by the system, 136 quintillion hashes per second (136 EH/s), in 2020. New blocks that document the recent transactions are confirmed and add to the blockchain only when a valid hash is found.

Simply put that the process of creation of BTC works in the following manner. The data of a transaction involving BTC are transmitted to all those participating in the (P2P) network, and so that the transaction can be made possible, it needs to be processed, or in other words, the cryptographic problem must be solved.

The miner receives 12.5 BTCs for each transaction block discovered, a payment for having loaned computational power to possibilitate the transactions in Bitcoin. The miners can also be rewarded with a fraction of the transactions done, which can be offered optionally by BTC users.

The rates are offered so that the transactions are prioritized by the miners in the formation of the candidate blocks, increasing the processing rate of the transaction (Camacho and da Silva, 2018).

2

Main Definitions

"Companies want you to be secure, but not against them."

Whitfield Diffie

This chapter presents the fundamental definitions to understand network security. It is based on the parameters established by the Internet Engineering Task Force (IETF) that publishes reports on protocol vulnerabilities, definitions given by the Center for the Study, Response and Treatment of Security Incidents in the Brazil (CERT.br), which is part of the Internet Steering Committee in Brazil, and by the Center for Internet Security (CIS), whose mission is to identify, develop, validate, promote, and sustain best practice solutions for cyber defense as well as build and lead communities to enable an environment of trust in cyberspace (CERT.br, 2019).

2.1 Criteria for a Security System

Claude E. Shannon, the famous cryptanalyst and information theorist, established some criteria to estimate the quality of a proposed security system (Shannon, 1949):

Security Quantity – There is a possibility to make a system perfectly secure, as in the case of the single-use cipher (one-time pad), in which the attacker does not obtain any information when intercepting a message. Other systems, while providing some information to the attacker, do not allow a single solution for intercepted cryptograms.

Key Dimension – The key must be transmitted by means that are not eligible for interception. Eventually, they need to be memorized. Therefore, the keys should be as short as possible.

19

Complexity for Encrypting and Decrypting – Encryption and decryption operations must be as simple as possible to minimize the time spent in the process, avoid errors, and also to reduce the complexity of the equipment involved.

Propagation of Errors – For some figures, an error in the encoding or transmission can lead to many errors in the clear text received, causing high loss of information. This can require frequent repetition of the encrypted message; therefore, the expansion of errors must be avoided.

Message Expansion – In some security systems, the message size is increased by the encryption process. This effect is undesirable and may appear when someone wants to hide the message statistics with the addition of nulls or when multiple substitutes are used.

Scientific proof of the security of the single-use cipher was carried out by Claude E. Shannon, who reported its results in a confidential report to the American Department of Defense, in 1945. The article with the results was eventually declassified and published in 1949 (Shannon, 1949).

2.2 Security Aspects to Consider

In the objective assessment of security aspects, it is important to consider for what purposes certain system should serve in order to guarantee the means for an adequate analysis of the type of threat to which the system is subjected (van der Lubbe, 1997).

The available means of security and preventive measures must be taken into account to avoid problems in that area. The dimension corresponds to the nature of the protection to be offered or whether security measures are in place for the prevention or correction of damages, possibly caused by security breaches. Figure 2.1 illustrates the relationship between the mentioned aspects.

Regarding the purpose of security measures, it is important to consider the confidentiality of data, which is important in the case of military or diplomatic matters. Messages exchanged between police bodies also require a degree of secrecy. The reliability of the information received, or its integrity, must also be maintained, as in the case of the banking system, which needs to guarantee the authenticity of an electronic transaction.

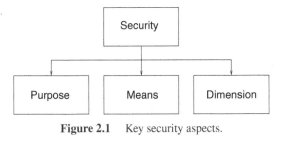

Figure 2.1   Key security aspects.

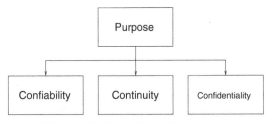

Figure 2.2   General security purposes.

Authentication of an interlocutor is important to validate his/her identity, for example. Furthermore, it is important that the system is protected against deliberate interruptions during transmission or data storage. Figure 2.2 shows the relationship between the mentioned points.

The means of ensuring security can be physical, when the system is protected against intrusion of strangers. This can be achieved with grids, safes, or intrusion sensors, for example. In any case, there is a need for organizational measures, because security measures should not be complicated or confusing for the users, under the risk of increasing the risks associated with negligence or carelessness with protection. It must be considered that human beings are liable to commit errors, and represent a weakness of any security system.

The measures discussed in this book are based on information theory and refer to algorithms and cryptographic methods, that is, they involve security with hardware actions, as in the case of protection with flash drives, or software, such as the security measures based on computer programs. Figure 2.3 illustrates the relationship between the mentioned items.

The security dimension may involve preventive, corrective, or damage-limiting measures. Encryption is generally used as a preventive measure to minimize the chance of occurring problems with the information transmitted.

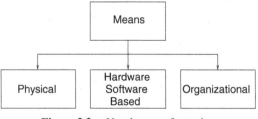

Figure 2.3   Usual types of security.

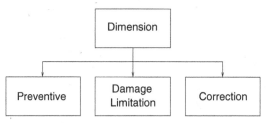

Figure 2.4   Typical security dimensions.

This protection results from the use of algorithms and protocols that are strong enough for a given application.

It must be taken into account that there are no absolutely safe measures and that the likelihood of security problems can only be minimized but never reduced to zero. Therefore, data reduction is an important measure in the event of an invasion of the system. Access must be kept limited, to prevent the attacker from obtaining all files, for example. In addition, furthermore, the system must be able to correct data that has been damaged. Figure 2.4 indicates how the mentioned points are related.

2.3 Malicious Codes

Malicious codes (malware) are programs designed to execute attacks on the operating system (OS) of a computer, causing damage or promoting harmful activities. There are several ways that malicious code can infect or compromise a computer (Stallings, 1999):

- automatic execution of infected removable devices, such as memory cards (memory stick or flash drive);
- direct attack, with computer invasion and inclusion of files with malicious codes;
- exploitation of existing vulnerabilities in programs installed on the computer;

- access to malicious Internet pages from vulnerable browsers;
- execution of previously infected files, obtained from electronic message attachments:
    - through removable devices;
    - by accessing unreliable website pages (on sites);
    - directly from other computers via shared network resources.

After installation, malicious programs gain access to data stored on the computer and can perform actions as if they were users, according to the permissions that each one has. Generally, the attacker develops and propagates malicious programs for financial gain, collect confidential information, self-promote, or vandalize specific sites.

Malicious programs are also used as intermediaries for the practice of crimes to conduct attacks and disseminate non-required information (spam). The main types of malicious code are presented in the following sections.

2.3.1 Virus

A virus is a program or part of a computer program, usually malicious, that spreads typically inserting copies of itself into other programs or files. In order to become active and continue the infection process, the virus depends on the execution of the program or host file, that is, for the computer to be infected, it is necessary to run a program that has been already infected.

The first viruses were written in machine language and only adhered to programs in time of execution. As this happened with each new execution of the program, eventually, the computer's memory was exhausted.

The virus spread through floppy disks. Because this type of media storage have become obsolete, new ways of dissemination have been developed, such as sending electronic message (e-mail). Currently, memory cards are the most widely used removable media for spreading viruses.

Viruses exhibit different behaviors. Some remain hidden, infecting files on the disk and performing activities without the user's knowledge. Others remain inactive for certain periods, entering activity on certain dates or from specific events. Some of the most common types of viruses are:

- Script virus – A program written in script language. This language allows the user to write programs for a special runtime environment that automates the execution of tasks. The virus can be passed by having access to a web page or by electronic mail, as an attached file or as part of the e-mail itself, written in Hypertext Markup Language (HTML). It can be run automatically, depending on the browser configuration and the user's message reader program.

- Macro virus – It is a specific type of virus from script, written in macro language, which is a programming language in which all computing is done by expanding (instantiating) macros, which means that standard inputs are eventually replaced by other output patterns. The virus aims to infect files manipulated by applications that use this language, such as Microsoft Office Suite components (Excel, Word, and PowerPoint, among others).
- Virus spread by electronic message – It is received as a file attached to a message, the content of which is intended to induce the user to trigger the file, causing it to be executed. When activated, it infects files and programs and sends its copies to the electronic addresses found in the lists of contacts stored on the computer.
- Cellphone virus – A virus that spreads between cell phones, through Bluetooth technology or by the use of the Multimedia Message Service (MMS). Infection occurs when a user authorizes the receipt of an infected file and executes it. After infecting the phone, the virus can destroy or overwrite files, remove or transmit contacts from the phone book, make phone calls, and drain the battery charge, in addition to trying to propagate to other cellphones.

2.3.2 Worms

The worm is a program, actually a process, capable of propagating itself automatically through networks, sending its copies to other computers.

Unlike the virus, a worm does not spread by including its copies in other programs or files but by directly executing its copies or by automatically exploiting existing vulnerabilities in programs installed on computers.

Worms consume a lot of computational resources, considering that they produce many copies of their own at the expense of intense computer use. As they spread, they can affect the performance of networks, in addition to using the computers.

The procedure for the propagation and infection of worms occurs as follows:

- Identification of target computers – After infecting a computer, the worm attempts to propagate and continue the infection process. To do this, it needs to identify the target computers for which it will try to copy itself, which can be done in the following ways:

- scan the network and identify active computers;
- wait for other computers to contact the infected computer;
- use lists, predefined or obtained from the Internet, containing the identification of targets;
- use information available on the infected computer, such as configuration files and lists of e-mail addresses.

- Sending copies – After identifying the targets, the worm makes its own copies and tries to send them for these computers, in one or more of the following ways:

  - from the exploitation of vulnerabilities in programs installed on the target computer;
  - attached to electronic messages;
  - through channels of Internet Relay Chat (IRC);
  - via instant messaging programs;
  - included in folders shared on local networks or of the type peer-to-peer (P2P).

- Activation of copies – After sending the copy, the worm needs to be executed in order for the infection to occur, which can happen in the following ways:

  - immediately after being transmitted, by exploiting vulnerabilities in programs that are running on the target computer at the time the copy is received;
  - directly by the user who can execute one of the copies sent to his computer;
  - by performing a specific user action, with which the worm is related, such as inserting removable media.

- Process restart – After the target computer is infected, the propagation and infection processes start again, and the target computer also begins to originate attacks.

2.3.3 Robots

The robot or bot is a program with communication mechanisms with the attacker that allow its remote control. The process of infection and propagation is similar to that of a worm, that is, it is able to propagate automatically, exploiting vulnerabilities in programs installed on computers.

Communication between the attacker and the computer infected by a bot can take place via an IRC, the network of servers that host channel or conversation rooms. Include web servers and network-type P2P, among other means. When communicating, the attacker can send instructions for malicious actions to be performed, such as launching attacks, stealing data from the infected computer and sending spam.

A computer infected with a bot is called a zombie because it can be controlled remotely, without the user's knowledge. It can also be called zombie spam when the bot installed turns it into a messaging server and uses it to send spam.

A botnet is a network made up of hundreds or thousands of zombie computers, which makes it possible to enhance the damaging actions of bots.

Typical malicious actions performed using botnets are denial of service attacks, propagation of malicious code, mass information gathering, sending spam, and camouflaging the identity of the attacker, using proxies installed on zombie computers.

2.3.4 Spy Programs

A spy program (spyware) is designed to monitor the activities of a system and send the collected information to third parties. It can be used in a legitimate or harmful way, depending on the installation procedure, on the performed actions, on the type of monitored information, and on the use that is made by whoever receives the collected information.

A spy program can be considered to be of legitimate use when installed on a personal computer, either by the owner himself or with his consent, in order to check if other people are using it in an abusive or unauthorized way.

However, it will be misused when performing actions that aim to compromise the user privacy or the computer security. This includes monitoring and capturing information regarding user navigation or that are inserted in other programs, such as, the user account and password.

Some specific types of spyware programs are as follows:

- Adware – Designed to display advertising, it can be used for legitimate purposes, when incorporated into programs and services, as an alternative to sponsorship or financial return for those who develop free programs or provide free services. And it can also be used for malicious purposes, when the advertisements presented are targeted according to how the user browses, without him knowing that such monitoring is being done. This program can make use of mobile computing and present options while the user walks, for example, in a shopping center.

- Keylogger – It is capable of capturing and storing the keys typed by the user on the computer keyboard, for eventual collection by an attacker. Its activation is sometimes subject to user action, such as access to a specific e-commerce or Internet site banking.
- Screenlogger – It is similar to keylogger and can store the position of the cursor and the content of the screen displayed on the monitor, when the mouse is triggered, or the region that surrounds the position where the mouse is clicked. It is used by attackers to capture the keys typed by users on virtual keyboards, available mainly on sites of Internet banking.

2.3.5 Backdoor Program

Backdoor is a program that allows an attacker to return to a compromised computer through the inclusion of services that are created or modified for this purpose. It can be included by the action of other malicious code, which has previously infected the computer, or by attackers, who exploit vulnerabilities in the programs installed on the computer to invade it.

Once included, the backdoor is used to ensure future access to the compromised computer, allowing it to have remote access, without the need to resort to the methods used in carrying out the invasion or infection and, generally, without being noticed.

The typical way of including a backdoor is to make a new service available or replace a given service for an altered version, usually with features that allow remote access. Remote administration programs, such as BackOrifice, NetBus, SubSeven, VNC, and Radmin, if they are poorly configured or used without the user's consent, can also be classified as backdoors.

There are backdoors included by program makers, who claim administrative needs. These cases pose a threat to the security of a computer. In addition to compromising user privacy, they can also be used by attackers to gain remote access to the computer.

2.3.6 Trojan Horse

Trojan horse, Trojan-horse, or just Trojan is a program that, in addition to performing the functions for which it was designed, also performs other functions, usually malicious, and without the user's knowledge.

Examples of Trojans are programs received or obtained from websites, which appear to be just animated virtual cards, photo albums, virtual games, and screen savers. These programs, in general, consist of a single file and need to be executed in order to be installed on the computer.

Trojans can also be installed by attackers who, after breaking into a computer, modify existing programs that, in addition to continuing to perform their original functions, also perform malicious actions.

There are different types of Trojan horses, classified according to the malicious actions that usually run when infecting a computer. Some of these types are:

- Trojan backdoor – Includes backdoors, allowing the attacker a remote access to the computer.
- Trojan banker – Collects bank details of the user, with the installation of spy programs that are activated when accessing Internet banking sites. It is similar to Trojan spy but with more specific objectives.
- Trojan clicker – Redirects the user's navigation to specific sites, in order to increase the number of accesses to these sites or to display advertisements.
- Destructive Trojan – Changes or deletes files and directories. It can format the hard drive and leave the computer inoperable.
- Trojan downloader – Installs other malicious code, obtained from websites.
- Trojan DoS – Installs denial of service tools and uses them to launch attacks.
- Trojan dropper – Installs other malicious code, and embeddeds in its own code.
- Trojan proxy – Installs a proxy server, allowing the computer to be used for anonymous browsing and sending spam.
- Trojan spy – Installs spy programs and uses them to collect sensitive information, such as passwords and credit card numbers, to send to the attacker.

2.3.7 Rootkit

The rootkit is a set of programs and techniques that allows you to hide and ensure the presence of an attacker, or other malicious code, on a compromised computer. The term rootkit comes from the combination of the words root, which corresponds to the superuser or administrator account of the computer on Unix or Linux systems, and kit, which indicates the set of programs used to maintain access privileges for this account.

The set of programs and techniques provided by rootkits can be used to:

- capture information from the network, on which the compromised computer is located, by intercepting traffic;
- hide activities and information, such as files, directories, processes, registry keys, and network connections;
- install other malicious code, such as backdoors, to ensure future access to the infected computer;
- map potential vulnerabilities on other computers by scanning the network;
- remove evidence in relevant event log files on a computer system logs.

The name rootkit does not indicate that the programs and the techniques that comprise it are used to obtain privileged access to a computer, but only to maintain it. Rootkits were used by attackers who, after breaking into a computer, installed them to keep the privileged access, without having to resort again to the methods used in the invasion, and to hide their activities from the person in charge or from the users of the computer.

2.4 Internet Scams

In general, it is not a simple task to attack and defraud data on a server of a banking or commercial institution, and, for this reason, scammers have been focusing their efforts on exploiting users' weaknesses. Using social engineering techniques and through different means and speeches, scammers seek to deceive and persuade potential victims to provide sensitive information or to take actions, such as executing malicious code and accessing false pages.

In possession of the victims' data, scammers usually carry out financial transactions, promote access to websites, send electronic messages, open ghost companies, and create illegitimate bank accounts, among other malicious activities. Many of the scams applied on the Internet can be considered crimes against property, typified as fraud.

The following sections present some of the main scams applied to the Internet and some precautions that must be taken to protect from them.

2.4.1 Identity Theft

Identity theft is the act by which a person tries to impersonate another person, by attributing a false identity in order to obtain undue advantages. Some cases of identity theft can be considered a crime against the public faith, typified as false identity.

From identity theft, for example, someone can open a business or bank account using the victim's name and documents. This can also happen on the Internet, with the creation of fake profiles on a social network, unauthorized access to the e-mail account, and sending messages posing as the victim.

The more information is made available on the Internet, the easier it is for a scammer to steal identities because the more data he/she has available and the more convincing he/she can be. In addition, the scammer may use other types of scams and attacks to collect information, including passwords, such as malicious code, hacking attacks brute force and traffic interception.

2.4.2 Resource Anticipation Fraud

Advance fee fraud is one in which a scammer seeks to induce a person to provide confidential information or to make an advance payment, with the promise of receiving some kind of benefit in the future.

By receiving electronic messages or accessing fraudulent websites, the person is involved in a situation that justifies sending personal information or making an advance payment in order to obtain a future benefit.

2.4.3 Phishing

Phishing, phishing-scam, or phishing/scam, is the type of fraud by which a scammer tries to obtain a user's personal and financial data, through the combined use of technical means and social engineering. It occurs through the sending of electronic messages that:

- they try to impersonate the official communication of a well-known institution, such as a bank, a company, or a popular site;
- they seek to attract the user's attention, whether out of curiosity, charity, or the possibility of obtaining some financial advantage;
- they inform that the non-execution of the described procedures can have serious consequences, such as enrollment in credit protection services and the cancelation of a bank account or a credit card registration;

- they try to induce the user to provide personal and financial data, by accessing false pages, which attempt to pass themselves off through the official website of the institution.

2.4.4 Pharming

Pharming is a specific type of phishing that involves redirecting user navigation to false websites, through changes made to the Domain Name System (DNS) service. In this case, when accessing a legitimate website, the browser is redirected, transparently, to a fake page. This redirection can occur:

- by compromising the DNS server of the provider used;
- by the action of malicious codes designed to alter the behavior of the computer's DNS service;
- by the direct action of an attacker, who may gain access to the computer's DNS service settings or modem with a high transmission rate.

2.4.5 E-commerce Scams

E-commerce scams are those in which the miscreants, in order to obtain financial advantages, exploit the relationship of trust existing between the parties involved in a commercial transaction. Some of these scams are presented below.

Fraudulent scam e-commerce site:
In this scam, the scammer creates a fraudulent website, with the specific objective of deceiving potential customers who, after making payments, do not receive the goods.

To increase the chances of success, the scammer often uses devices such as: send spam, advertise via sponsored links, announce discounts on collective shopping sites, and offer highly sought-after products with prices below those practiced by the market.

Scam involving collective shopping sites
Collective shopping sites have been used in e-commerce scams. In addition to the risks inherent in day-to-day business relationships, collective purchasing sites also have their own risks, generated mainly by the pressure placed on consumers to make quick decisions so as not to miss the opportunity to buy.

Scammers create fraudulent websites and use them to advertise products on collective shopping sites and thus reach large numbers of victims in a short period of time.

In addition, collective shopping sites can also be used as the subject of phishing messages. Scammers usually send messages as if they originated on the real site and, in this way, try to induce the user to access a fake page and provide personal data, such as credit card number and passwords.

Auction site and product sale scam

In the scam of the site of auction and sale of products, a buyer or seller acts in bad faith and does not comply with the agreed obligations or uses the personal and financial data involved in the commercial transaction for other purposes. For example:

- The buyer tries to receive the goods without making the payment or does so by transferring from an illegitimate or stolen bank account.
- The seller tries to receive payment without delivering the goods or the damaged, counterfeit delivery, with characteristics different from those advertised or acquired in an illegal and criminal manner.
- The buyer, or the seller, sends false messages, on behalf of the payment management system, as a way of proving that the payment has been made or that the goods have not been shipped.

2.4.6 Hoax

A hoax is a message that has alarming or false content and that generally has as its sender, or points out as an author, some institution, important company, or government agency. Through a thorough reading of its content, it is usually possible to identify meaningless information and attempted strikes, such as chains and pyramids.

Hoaxes can bring several problems, both for those who receive and distribute them, and for those who are cited in their content. Among other problems, a hoax can:

- contain malicious programs;
- spread misinformation on the Internet;
- needlessly occupy space in users' e-mail boxes;

- compromise the credibility and reputation of people, or entities, mentioned in the message;
- compromise the credibility and reputation of the person who passes it on;
- increase the load on messaging servers and traffic on the network;
- indicate, in the message content, actions to be taken and which can result in serious damage, such as deleting a file that supposedly contains malicious code but is part of the computer's operating system.

2.5 Internet Attacks

Attacks tend to occur on the Internet for different purposes, aiming at different targets and using different techniques. Any service, computer, or network that is accessible via the Internet can be the target of an attack, just as any computer with access to the Internet can participate in an attack.

The reasons that lead attackers to launch attacks on the Internet are diverse, ranging from simple amusement to criminal actions. Some examples are as follows

- Demonstration of power – Show a company that it can be invaded or have its services suspended and, thus, try to sell services or blackmail it so that the attack does not occur again.
- Prestige – To boast, before other attackers, for having managed to break into computers, render services inaccessible or disfigure sites considered targeted or difficult to be attacked. Dispute with other attackers who can carry out the greatest number of attacks or be the first to reach a certain target.
- Financial motivations – Collect and use confidential information from users to apply scams.
- Ideological motivations – Make inaccessible or invade sites that disseminate content contrary to the attacker's opinion or disseminate messages of support or contrary to a certain ideology.
- Business motivations – Make inaccessible or invade sites and computers of competing companies, in an attempt to prevent customers from accessing or compromising the reputation of these companies.

2.5.1 Vulnerability Exploitation

A vulnerability is defined as a condition that, when exploited by an attacker, could result in a security breach. Examples of vulnerabilities are flaws in the design, implementation or configuration of programs, services, or network equipment.

A vulnerability scanning attack occurs when an attacker, using a vulnerability, attempts to perform malicious actions, such as breaking into a system, accessing sensitive information, launching attacks on other computers, or making a service inaccessible.

2.5.2 Network Scan

Network scanning consists of making detailed searches on networks, with the objective of identifying active computers and collecting information about them, such as available services and installed programs. Based on the information collected, it is possible to associate possible vulnerabilities to the available services and the programs installed on the detected active computers.

Network scanning and exploiting associated vulnerabilities can be used in the following ways:

- Legitimate – By duly authorized persons, to check the security of computers and networks and, thus, take corrective and preventive measures.
- Malicious – By attackers, to exploit the vulnerabilities found in the services provided and in the programs installed to perform malicious activities. Attackers can also use detected active computers as potential targets in the process of automatically spreading malicious code and in brute force attacks.

2.5.3 Fake E-mail Address

E-mail spoofing is a technique that consists of altering the header fields of an e-mail in order to make it appear that it was sent from a certain source when, in fact, it was sent from other.

This is possible due to the characteristics of the Simple Mail Transfer Protocol (SMTP) that allows header fields, such as "From:" (address of the person who sent the message), "Reply-To" (reply address of the message), and "Return-Path" (address where possible errors in sending the message are reported), are falsified.

Such attacks are used for the propagation of malicious code, sending of spam, and in phishing scams. Attackers use e-mail addresses collected from infected computers to send messages and try to make their recipients believe that they came from people they know.

2.5.4 Traffic Interception

Traffic interception (sniffing) consists of inspecting data transmitted on computer networks by the use of specific programs called sniffers. This technique can be used in a way that is:

- Legitimate – By network administrators, to detect problems, analyze performance, and monitor malicious activities related to the computers or networks managed by them.
- Malicious – By attackers, to capture sensitive information, such as passwords, credit card numbers, and the content of confidential files that are traveling through insecure connections, that is, without encryption.

2.5.5 Brute Force Attack

A brute force attack consists of searching, by trial and error, for a username and password and, thus, executing processes and accessing websites, computers, and services in the name and with the same privileges as this user.

Any computer, network equipment, or service that is accessible via the Internet, with a username and password, can be the target of a brute force attack. Mobile devices, which are password protected, in addition to being able to be attacked by the network, can also be the target of this type of attack if the attacker has physical access to them.

If an attacker is aware of the target username and password, he or she can perform malicious actions on his/her behalf, such as:

- change the password, making it difficult for the user to access the hacked website or computer;
- invade the e-mail service used and have access to the content of messages and the contact list, in addition to being able to send messages on behalf of the user;
- having access to the victim's social network and sending messages to his followers containing malicious codes or changing his privacy options;

- hack the computer and, according to the user's permissions, perform actions such as deleting files, obtaining confidential information, and installing malicious code.

2.5.6 Page Defacement

Page defacement or graffiti is a technique that consists of altering the content of a website's Web page. The main ways that an attacker, in this case calso called a defacer, can use to deface a Web page are:

- explore application errors;
- exploit application server vulnerabilities;
- explore vulnerabilities in the programming language or packages used in the development of the Web application;
- invade the server where the application on web is hosted and directly change the files that make up the site;
- stealing passwords for accessing the Web interface used for remote administration.

2.5.7 Denial of Service

Denial of service (DoS) is a technique by which an attacker uses a computer to shut down a service, a computer, or a network connected to the Internet. When used in a coordinated and distributed manner, that is, when a set of computers is used in the attack, it is called a distributed denial of service or DDoS.

The purpose of these attacks is not to invade or collect information but to deplete resources and cause unavailability to the target. When this occurs, all people who depend on the affected resources are harmed, as they are unable to access or perform the desired operations.

In the already registered cases of attacks, the targets were prevented from offering services during the period in which they occurred, but, in the end, they returned to operating normally, without any leakage of information or compromise of systems or computers.

A person can voluntarily use tools and have his computer used in attacks. Most computers, however, participate in attacks without the knowledge of their owner, because they are infected and are part of botnets.

Denial of service attacks can be carried out by various means, such as:

- by sending a large number of requests for a service, consuming the resources necessary for its operation (processing, number of

simultaneous connections, memory, and disk space, for example) and preventing requests from other users from being met;
- for the generation of high data traffic for a network, depleting the available bandwidth and making any access to computers or services on this network unavailable;
- for the exploitation of existing vulnerabilities in programs, which can make a particular service inaccessible.

In situations where there is a saturation of resources, if a service has not been well dimensioned, it may be inoperable when trying to meet its own legitimate requests.

2.6 Cryptography

Encryption, considered the science and art of writing messages in encrypted form or in code, is one of the main security mechanisms a user can utilize to be protected from the risks associated with using the Internet.

At first glance, it may seem complicated, but in order to enjoy the benefits it provides, there is no need to study the subject or be an experienced mathematician. Currently, encryption is already integrated, or can be added, to most operating systems and applications, and to use it, many times, it is enough to perform some settings or mouse clicks.

With the use of encryption it is possible to:

- protect sensitive data stored on the computer, such as the password file or work performed;
- create a specific area (partition) on the computer, in which all the information recorded there will be automatically encrypted;
- protect backups against unauthorized access, especially those sent to external media storage areas;
- protect communications carried out over the Internet, such as messages sent or received and banking and commercial transactions carried out.

2.6.1 Symmetric Key and Asymmetric Key Cryptography

Depending on the type of key used, cryptographic methods can be subdivided into two broad categories: symmetric key cryptography and asymmetric key cryptography.

Symmetric key cryptography, also called secret or single-key cryptography, uses the same key to both encrypt and decrypt information and is used primarily to ensure data confidentiality. In cases where the information is encrypted and decoded by the same person, there is no need to share the secret key.

However, when these operations involve different people or equipment, it is necessary that the secret key be previously combined through a secure communication channel. Examples of cryptographic methods that use a symmetric key are: DES, AES, Blowfish, RC4, 3DES, and IDEA.

Asymmetric key cryptography, also known as public key cryptography, uses two distinct keys: a public one, which can be freely disclosed, and a private one, which must be kept secret. When information is encrypted with one of the keys, only the other key in the pair can decode it.

Which key to use to encrypt depends on the specific protection, whether confidentiality or authentication, integrity and non-repudiation. The private key can be stored in different ways, such as a file on the computer, a smart card, or a token. Examples of cryptographic methods that use asymmetric keys are: RSA, DSA, ECC, and Diffie-Hellman.

Symmetric key cryptography, when compared to asymmetric keys, is the best way to guarantee the confidentiality of large volumes of data, as it is faster to process. However, when used for information sharing, it becomes complex and poorly scalable, due to:

- need for a secure communication channel to promote the sharing of the secret key between the parties;
- difficulty in managing large numbers of keys.

The encryption of asymmetric keys, despite having a slower processing than that of the symmetric key, solves these problems since it facilitates management, as it does not require that a secret key be kept with each one who wishes to communicate and dispenses with the need to a secure communication channel for key sharing.

To take advantage of these methods, the combined use of both is ideal. Symmetric key cryptography is used to encrypt information and asymmetric key cryptography is used to share the secret key, also called the session key. Combined usage is what is used by Web browsers and e-mail reader programs. Examples of using this combined method are: SSL, PGP, and secure/multipurpose Internet mail extensions (S/MIME).

2.6.2 Hash Function

A hash function is a cryptographic method that, when applied to information, regardless of its size, generates a single, fixed-size result, called hash.

A hash can be used to:

- Check the integrity of a file stored on the computer or in backups.
- Check the integrity of a file obtained from the Internet. Some sites provide the corresponding hash so that it is possible to verify that the file was correctly transmitted and recorded.
- Generate digital signatures.

To check the integrity of a file, just calculate its hash. When necessary, this code can be generated again. If the two hashes are equal, then it is possible to conclude that the file has not been changed. Otherwise, it may be an indication that the file is corrupted or that it has been modified. Examples of hashing methods are: SHA-1, SHA-256, and MD5.

2.6.3 Digital Signature

The digital signature allows verifying the authenticity and integrity of an information, that is, that it was actually generated by whoever claims to have done this and that it has not been altered.

The digital signature is based on the fact that only the owner knows the private key and that if it was used to encrypt the information, then only its owner could have done this. The signature is verified using the public key because if the text has been encrypted with the private key, only the corresponding public key can decrypt it.

To circumvent the low efficiency characteristic of asymmetric key encryption, the hash is encoded and does not give any information about the contents of the file.

2.6.4 Digital Certificate

The public key can be freely disclosed. However, if there is no way to prove who it belongs to, the user can communicate, in encrypted form, with an impostor.

An impostor can create a fake public key for a friend of the target and send it or make it available in a repository. By using it to encrypt information for his friend, the user is actually encrypting it for the impostor, who has the corresponding private key and will be able to decrypt it. One way to prevent this from happening is by the use of digital certificates.

The digital certificate is an electronic record composed of a set of data that distinguishes an entity and associates it with a public key. It can be issued to people, companies, equipment, or services on the network (an Internet site) and can be approved for different uses, such as confidentiality and digital signature.

A digital certificate can be compared to an identity document, which contains the user's personal data and the identification of who issued it. In the case of the digital certificate, the entity is a certification authority (CA).

An issuing CA is also responsible for publishing information about certificates that are no longer trusted. Whenever the CA discovers or is informed that a certificate is no longer trusted, it includes it in a black list, called the certificate revocation list (CRL) for users to be aware of. The CRL is an electronic file published periodically by the CA, containing the serial number of the certificates that are no longer valid and the date of revocation.

The fields of the digital certificate are standardized, but the graphical representation may vary between different browsers and operating systems. In general, the basic data that make up a digital certificate are:

- version and serial number of the certificate;
- data that identifies the CA that issued the certificate;
- data that identifies to whom the certificate was issued;
- public key of the certificate owner;
- validity of the certificate, indicating when it was issued and when it is valid;
- digital signature of the issuing CA and data for verification of the signature.

A CA's digital certificate is usually issued by another CA, establishing a hierarchy known as the certificate chain or certification path. The root CA, the chain's first authority, is the anchor of trust for the entire hierarchy and, because there is no other CA above it, it has a self-signed certificate.

Publicly recognized root CA certificates are already included, by default, in most operating systems and browsers and are updated along with the systems themselves. Some examples of updates made to the browsers' certificate base are: adding new CA's, renewing expired certificates, and deleting untrusted CA's.

Some special types of digital certificate are as follows:

- Self-signed certificate – One in which the owner and the issuer are the same entity. It is usually used in two ways:

    - Legitimate – In addition to the root CAs, self-signed certificates are also often used by educational institutions and small groups that want to provide confidentiality and integrity in connections but who do not want to, or cannot, bear the burden of acquiring a digital certificate validated by a commercial CA.

 – Malicious – An attacker can create a self-signed certificate and use, for example, e-mail messages phishing to trick users into installing it. From the moment the certificate is installed in the browser, it will be possible to establish encrypted connections with fraudulent sites, without the browser issuing alerts about the certificate's reliability.

• Extended Validation Certificate Secure Socket Layer (EV SSL) – Certificate issued under a more rigorous validation process for the requester. It includes verifying that the company has been legally registered, is active, and that it holds the domain registration for which the certificate will be issued, in addition to additional data, such as the physical address.

2.6.5 Cryptography Programs

To ensure the security of the messages, it is important to use e-mail reader programs with native encryption support, such as those that implement S/MIME, or that allow the integration of other specific programs and add-ons for this purpose.

Encryption programs, such as GnuPG2, in addition to being able to be integrated with programs that read e-mails, can also be used separately to encrypt other types of information, such as files stored on the computer or on removable media.

There are also programs native to the operating system or purchased separately that allow to encrypt the entire computer disk, file directories, and external storage devices, such as memory cards and hard drives, which aim to preserve the confidentiality of information in case of loss or theft of equipment.

3

Information Theory

"Few persons can be made to believe that it is not quite an easy thing to invent a method of secret writing which shall baffle investigation."

Edgar Allan Poe; A Few Words On Secret Writing, 1841

Information theory is a branch of probability theory which has application and correlation with many areas, including communication systems, communication theory, physics, language and meaning, cybernetics, psychology, art, and complexity theory (Pierce, 1980). The basis for the theory was established by Harry Theodor Nyquist (1889–1976) (Nyquist, 1924), also known as Harry Nyquist, and Ralph Vinton Lyon Hartley (1888–1970), who invented the Hartley oscillator (Hartley, 1928). They published the first articles on the subject, in which the factors that influenced the transmission of information were discussed.

The seminal article by Claude E. Shannon (1916–2001) extended the theory to include new factors, such as the noise effect in the channel and the savings that could be obtained as a function of the statistical structure of the original message and the information receiver characteristics (Shannon, 1948b). Shannon defined the fundamental communication problem as the possibility of, exactly or approximately, reproducing, at a certain point, a message that has been chosen at another one.

The main semantic aspects of the communication, initially established by Charles Sanders Peirce (1839–1914), a philosopher and creator of semiotic theory, are not relevant for the development of the Shannon information theory. What is important is to consider that a particular message is selected from a set of possible messages.

Of course, as mentioned by John Robinson Pierce (1910–2002), quoting the philosopher Alfred Jules Ayer (1910–1989), it is possible to communicate not only information but also knowledge, errors, opinions, ideas, experiences, desires, commands, emotions, and feelings. Heat and movement can be communicated, as well as force, weakness, and disease (Pierce, 1980).

Hartley has found several reasons as to why the natural information should measure the logarithm:

- It is a practical metric in engineering, considering that various parameters, such as time and bandwidth, are proportional to the logarithm of the number of possibilities.
- From a mathematical point of view, it is an adequate measure because several limit operations are simply stated in terms of logarithms.
- It has an intuitive appeal, as an adequate metric, because, for instance, two binary symbols have four possibilities of occurrence.

The choice of the logarithm base defines the information unit. If base 2 is used, the unit is the bit, an acronym suggested by John W. Tukey for binary digit, that is also a play or words, that can also mean a piece of information. The information transmission is informally given in bit/s, but a unit has been proposed to pay tribute to the scientist who developed the concept; it is called the shannon or [Sh] for short. This has a direct correspondence with the unit for frequency, hertz or [Hz], for cycles per second, which was adopted by the International System of Units (SI)[1].

Aleksandr Yakovlevich Khinchin (1894–1959) (Khinchin, 1957) put the information theory in solid basis, with a more precise and unified mathematical discussion about the entropy concept, which supported Shannon's intuitive and practical view.

The books by Robert B. Ash (Ash, 1965) and Amiel Feinstein (Feinstein, 1958) give the mathematical reasons for the choice of the logarithm to measure information, and the book by J. Aczél e Z. Daróczy (Aczél and Daróczy, 1975) presents several Shannon information measures and their characterization as well as Alfréd Rényi (1921–1970) entropy metric.

A discussion on generalized entropies can be found in the book edited by Luigi M. Ricciardi (Ricciardi, 1990). Lotfi Asker Zadeh introduced the concept of fuzzy set, an efficient tool to represent the behavior of systems that depend on the perception and judgment of human beings, and applied it to information measurement (Zadeh, 1965).

---

[1]The author of this book proposed the adoption of the shannon [Sh] unit during the IEEE International Conference on Communications (ICC'2001), in Helsinki, Finland, shortly after Shannon's death.

3.1 Information Measurement

The objective of this section is to establish a measure for the information content of a discrete system, using probability theory. Consider a discrete random experiment, such as the occurrence of a symbol, and its associated sample space Ω, in which X is a real random variable (Reza, 1961).

The random variable X can assume the following values:

$$X = \{x_1, x_2, \ldots, x_n\},$$

$$\text{in which} \bigcup_{k=1}^{N} x_k = \Omega, \tag{3.1}$$

with probabilities in the set P

$$P = \{p_1, p_2, \ldots, p_n\},$$

$$\text{in which} \sum_{k=1}^{N} p_k = 1. \tag{3.2}$$

The information associated with a particular event is given by

$$I(x_i) = \log\left(\frac{1}{p_i}\right), \tag{3.3}$$

which is meaningful because the sure event has probability one and zero information, by a property of the logarithm, and the impossible event has zero probability and infinite information.

Example: Suppose the sample space is partitioned into two equally probable spaces. Then

$$I(x_1) = I(x_2) = -\log \tfrac{1}{2} = 1 \text{ bit.} \tag{3.4}$$

That is, the choice between two equally probable events requires one unit of information, when a base 2 logarithm is used.

Considering the occurrence of 2^N equiprobable symbols, the self-information of each event is given by

$$I(x_k) = -\log p_k = -\log 2^{-N} = N \text{ bits.} \tag{3.5}$$

Table 3.1   Symbol probabilities of a two-symbol source.

Symbol	Probability
x_1	$\frac{1}{4}$
x_2	$\frac{3}{4}$

It is possible to define the source entropy, $H(X)$, as the average information, obtained by weighing of all the occurrences

$$H(X) \;=\; E[I(x_i)] \;=\; -\sum_{i=1}^{N} p_i \log p_i. \tag{3.6}$$

Observe that eqn 3.6 is the weighing average of the logarithms of the probabilities, in which the weights are the real values of the probabilities of the random variable X, and this indicates that $H(X)$ can be interpreted as the expected value of the random variable that assumes the value $-\log p_i$, with probability p_i (Ash, 1965).

Example: Consider a source that emits two symbols, with unequal probabilities, given in Table 3.1.

The source entropy is calculated as

$$H(X) \;=\; -\frac{1}{4}\log\frac{1}{4} \;-\; \frac{3}{4}\log\frac{3}{4} = 0.81 \text{ bits per symbol.}$$

3.2 Requirements for an Information Metric

A few fundamental properties are necessary for the entropy in order to obtain an axiomatic approach to base the information measurement (Reza, 1961).

- If the event probabilities suffer a small change, the associated measure must change in accordance, in a continuous manner, which provides a physical meaning to the metric

$$H(p_1, p_2, \;\cdots\; , p_N) \text{ is continuous in } p_k, k \;=\; 1, \, 2, \;\cdots\; , \, N, \tag{3.7}$$
$$0 \le p_k \le 1.$$

- The information measure must be symmetric in relation to the probability set P. The is, the entropy is invariant to the order of events

$$H(p_1, p_2, p_3, \;\cdots\; , p_N) \;=\; H(p_1, p_3, p_2, \;\cdots\; , p_N). \tag{3.8}$$

- The maximum of the entropy is obtained when the events are equally probable. That is, when nothing is known about the set of events, or about what message has been produced, the assumption of a uniform distribution gives the highest information quantity that corresponds to the highest level of uncertainty

$$\text{Maximum of } H(p_1, p_2, \ldots, p_N) = H\left(\frac{1}{N}, \frac{1}{N}, \ldots, \frac{1}{N}\right). \quad (3.9)$$

Example: Consider two sources that emit four symbols. The first source symbols, shown in Table 3.2, have equal probabilities, and the second source symbols, shown in Table 3.3, are produced with unequal probabilities.

The mentioned property indicates that the first source attains the highest level of uncertainty, regardless of the probability values of the second source, as long as they are different.

- Consider that an adequate measure for the average uncertainty has been found $H(p_1, p_2, \ldots, p_N)$ associated with a set of events. Assume that event $\{x_N\}$ is divided into M disjoint sets, with probabilities q_k, such that

$$p_N = \sum_{k=1}^{M} q_k, \quad (3.10)$$

Table 3.2   Identically distributed symbol probabilities.

Symbol	Probability
x_1	$\frac{1}{4}$
x_2	$\frac{1}{4}$
x_3	$\frac{1}{4}$
x_4	$\frac{1}{4}$

Table 3.3   Unequal symbol probabilities.

Symbol	Probability
x_1	$\frac{1}{2}$
x_2	$\frac{1}{4}$
x_3	$\frac{1}{8}$
x_4	$\frac{1}{8}$

and the probabilities associated with the new events can be normalized in such a way that

$$\frac{q_1}{p_n} + \frac{q_2}{p_n} + \cdots + \frac{q_m}{p_n} = 1. \tag{3.11}$$

Then, the creation of new events from the original set modifies the entropy to

$$H(p_1,p_2, \ldots, p_{N-1},q_1,q_2, \ldots, q_M) = H(p_1,\ldots,p_{N-1},p_N)$$
$$+ \quad p_N H\left(\frac{q_1}{p_N}, \frac{q_2}{p_N}, \ldots, \frac{q_M}{p_N}\right), \tag{3.12}$$

with

$$p_N = \sum_{k=1}^{M} q_k.$$

- Finally, adding or removing an event with probability zero does not contribute to the entropy,

$$H_{n+1}(p_1,\ldots,p_n,0) = H_n(p_1,\ldots,p_n). \tag{3.13}$$

It is possible to show that the function defined by eqn 3.6 satisfies all requirements. To demonstrate the continuity, it suffices to do (Reza, 1961)

$$H(p_1,p_2,\ldots,p_N) = -[p_1\log p_1 + p_2\log p_2 + \cdots + p_N\log p_N]$$
$$= -[p_1\log p_1 + p_2\log p_2 + \cdots p_{N-1}\log p_{N-1}$$
$$+ (1 - p_1 - p_2 - \cdots - p_{N-1})$$
$$\cdot \ \log(1 - p_1 - p_2 - \cdots - p_{N-1})]. \tag{3.14}$$

Note that, for all independent random variables, the complete set of probabilities $\{p_1,p_2,\ldots,p_{N-1}\}$ and also $(1 - p_1 - p_2 - \ldots - p_{N-1})$ are contiguous in $[0,1]$ and that the logarithm of a continuous function is also continuous. The entropy is clearly symmetric.

The maximum value property can be demonstrated, if one considers that all probabilities are equal and that the entropy is maximized by that condition

$$p_1 = p_2 = \cdots = p_N. \tag{3.15}$$

Taking into account, according to intuition, that the uncertainty is maximum for a system of equiprobable states, it is possible to arbitrarily choose a random variable with probability p_N depending on p_k, and $k = 1,2,\ldots, N - 1$. Taking the derivative of the entropy in terms of each probability

$$\frac{dH}{dp_k} = \sum_{i=1}^{N} \frac{\partial H}{\partial p_i} \frac{\partial p_i}{\partial p_k}$$

$$= -\frac{d}{dp_k}(p_k \log p_k) - \frac{d}{dp_N}(p_N \log p_N) \frac{\partial p_N}{\partial p_k}. \qquad (3.16)$$

But, probability p_N can be written as

$$p_N = 1 - (p_1 + p_2 + \cdots + p_k + \cdots + p_{N-1}). \qquad (3.17)$$

Therefore, the derivative of the entropy is

$$\frac{dH}{dp_k} = -(\log_2 e + \log p_k) + (\log_2 e + \log p_n), \qquad (3.18)$$

which, using a property of logarithms, simplifies to

$$\frac{dH}{dp_k} = -\log \frac{p_k}{p_n}. \qquad (3.19)$$

But

$$\frac{dH}{dp_k} = 0, \text{ which gives } p_k = p_N. \qquad (3.20)$$

As p_k was chosen in an arbitrary manner, one concludes that, to obtain a maximum for the entropy function, one must have

$$p_1 = p_2 = \cdots = p_N = \frac{1}{N}. \qquad (3.21)$$

The maximum is guaranteed because

$$H(1,0,0, \ldots ,0) = 0. \qquad (3.22)$$

On the other hand, for equiprobable events, it is possible to verify that the entropy is always positive, for it attains its maximum at (Csiszár and Kórner, 1981)

$$H\left(\frac{1}{N},\frac{1}{N}, \cdots ,\frac{1}{N}\right) = \log N > 0. \tag{3.23}$$

To prove additivity, it suffices to use the definition of entropy, computed for a two-set partition, with probabilities $\{p_1, p_2, \ldots, p_{N-1}\}$ e $\{q_1, q_2, \ldots, q_M\}$

$$\begin{aligned}
H(p_1,p_2, \cdots ,p_{N-1},q_1,q_2, \cdots ,q_M) &= \\
&= -\sum_{k=1}^{N-1} p_k \log p_k - \sum_{k=1}^{M} q_k \log q_k \\
&= -\sum_{k=1}^{N} p_k \log p_k + p_N \log p_N - \sum_{k=1}^{M} q_k \log q_k \\
&= H(p_1,p_2, \cdots ,p_N) + p_N \log p_N \\
&\quad - \sum_{k=1}^{M} q_k \log q_k.
\end{aligned} \tag{3.24}$$

But, the second part of the last term can be written in a way to display the importance of the entropy in the derivation

$$\begin{aligned}
p_N \log p_N - \sum_{k=1}^{M} q_k \log q_k &= p_N \sum_{k=1}^{M} \frac{q_k}{p_N} \log p_N - \sum_{k=1}^{M} q_k \log q_k \\
&= -p_N \sum_{k=1}^{M} \frac{q_k}{p_N} \log \frac{q_k}{p_N} \\
&= p_N H\left(\frac{q_1}{p_N},\frac{q_2}{p_N}, \cdots , \frac{q_M}{p_N}\right),
\end{aligned} \tag{3.25}$$

and this demonstrates the mentioned property.

The entropy is non-negative, which guarantees that the partitioning of one event into several other events does not reduce the system entropy, as shown in the following:

$$H(p_1,\ldots,p_{N-1},q_1,q_2,\ldots,q_M) \geq H(p_1,\ldots,p_{N-1},p_N). \tag{3.26}$$

That is, if one splits a symbol into two or more, the entropy always increases, and that is the physical origin of the word.

Example: Consider a binary source, X, that emits symbols 0 and 1 with probabilities p and $q = 1 - p$. The average information per symbol is given by $H(X) = -p \log p - q \log q$, which is known as entropy function

$$H(p) = -p \log p - (1 - p) \log (1 - p). \tag{3.27}$$

Example: For the binary source, consider that the symbol probabilities are $p = 1/8$ and $q = 7/8$, and compute the entropy of the source.

The average information per symbol is given by

$$H(X) = -1/8 \log 1/8 - 7/8 \log 7/8,$$

which gives $H(X) = 0.544$.

Note that even though 1 bit is produced for each symbol, the actual average information is 0.544 bits due to the unequal probabilities.

The entropy function has a maximum when all symbols are equiprobable, for $p = q = 1/2$, for which the entropy is 1 bit/symbol. The function attains a minimum for $p = 0$ or $p = 1$.

This function plays an essential role in determining the capacity of a binary symmetric channel. Observe that the entropy function is concave, that is,

$$H(p_1) + H(p_2) \leq 2H \left(\frac{p_1 + p_2}{2} \right). \tag{3.28}$$

The entropy function is illustrated in Figure 3.1, in which it is possible to notice the symmetry, concavity, and the maximum for equiprobable symbols. As a consequence of the symmetry, the sample spaces, with probability distributions obtained from permutations of a common distribution, provide the same information quantity (van der Lubbe, 1997).

Example: Consider a certain source that emits symbols from a given alphabet $X = \{x_1, x_2, x_3, x_4\}$, with probabilities given in Table 3.4. What is the entropy of this source?

The entropy is computed using Formula(3.6), for $N = 4$ symbols, as

$$H(X) = - \sum_{i=1}^{4} p_i \log p_i$$

or

$$H(X) = -\frac{1}{2} \log \frac{1}{2} - \frac{1}{4} \log \frac{1}{4} - \frac{2}{8} \log \frac{1}{8} = 1.75 \text{ bits per symbol.}$$

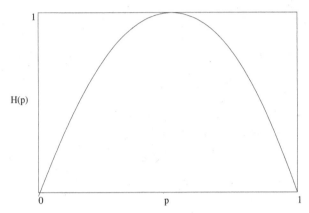

Figure 3.1   Graph of an information function.

Table 3.4   Symbol probabilities of a certain source.

Symbol	Probability
x_1	$\frac{1}{2}$
x_2	$\frac{1}{4}$
x_3	$\frac{1}{8}$
x_4	$\frac{1}{8}$

3.3 Source Coding

The efficient representation of data produced by a discrete source is called source coding. For a source coder to obtain a good performance, it is necessary to take the symbol statistics into account. If the symbol probabilities are different, it is useful to assign short codewords to probable symbols and long ones to infrequent symbols. This produces a variable length code, such as the Morse code.

Two usual requirements to build an efficient code are:

1. The codewords generated by the coder are binary.
2. The codewords are unequivocally decodable, and the original message sequence can be reconstructed from the binary coded sequence.

Consider Figure 3.2, which shows a memoryless discrete source, whose output x_k is converted by the source coder into a sequence of 0s and 1s, denoted b_k. Assume that the source alphabet has K different symbol and that the k-ary symbol, x_k, occurs with probability p_k, $k = 0, 1, \ldots, K - 1$.

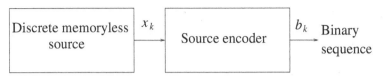

Figure 3.2 Source encoder.

Let l_k be the average length, measured in bits, of the binary word assigned to symbol x_k. The average length of the words produced by the source coder is defined as (Haykin, 1988)

$$\overline{L} = \sum_{k=1}^{K} p_k l_k. \tag{3.29}$$

The parameter $\overline{L}$ represents the average number of bits per symbol from those that are used in the source coding process. Let $L_{\min}$ be the smallest possible value of $\overline{L}$. The source coding efficiency is defined as (Haykin, 1988)

$$\eta = \frac{L_{\min}}{\overline{L}}. \tag{3.30}$$

Because $\overline{L} \geq L_{\min}$, then $\eta \leq 1$. The source coding efficiency increases as η approaches 1.

Shannon's first theorem, or source coding theorem, provides a means to determine $L_{\min}$ (Haykin, 1988).

> Given a memoryless discrete source with entropy $H(X)$, the average length of the codewords is limited by
>
> $$\overline{L} \geq H(X).$$

Entropy $H(X)$, therefore, represents a fundamental limit for the average number of bits per source symbol $\overline{L}$, that are needed to represent a memoryless discrete source, and this number can be as small as, but never smaller than, the source entropy $H(X)$. Therefore, for $L_{\min} = H(X)$, the source coding efficiency can be written as (Haykin, 1988)

$$\eta = \frac{H(X)}{\overline{L}}. \tag{3.31}$$

The code redundancy is given by (Abramson, 1963)

$$1 - \eta = \frac{\overline{L} - H(X)}{\overline{L}}. \tag{3.32}$$

3.4 Extension of a Memoryless Discrete Source

It is useful to consider the encoding of blocks of N successive symbols from the source, instead of individual symbols. Each block can be seen as a product of an extended source, with an alphabet X^N that has K^N distinct blocks. The symbols are statistically independent; therefore, the probability of an extended symbol is the product of the probabilities of the original symbols, and it can be shown that

$$H(X^N) = NH(X). \tag{3.33}$$

Example: Consider the discrete memoryless source with alphabet

$$X = \{x_1, x_2, x_3\}.$$

The second-order extended source has an alphabet

$$X^2 = \{x_1x_1, x_1x_2, x_1x_3, x_2x_1, x_2x_2, x_2x_3, x_3x_1, x_3x_2, x_3x_3\}.$$

For the second-order extended source of the example, $p(x_i x_j) = p(x_i)p(x_j)$. In particular, if all original source symbols are equiprobable, then $H(X) = \log_2 3$ bits. The second-order extended source has nine equiprobable symbols; therefore, $H(X^2) = \log_2 9$ bits. It can be noticed that $H(X^2) = 2H(X)$.

3.4.1 Improving the Coding Efficiency

The following example illustrates how to improve the coding efficiency using extensions of a source (Abramson, 1963).

Example: Consider a memoryless source $S = \{x_1, x_2\}$, with $p(x_1) = \frac{3}{4}$ and $p(x_2) = \frac{1}{4}$. The source entropy is given by $\frac{1}{4}\log_2 4 + \frac{3}{4}\log_2 \frac{4}{3} = 0.811$ bits.
A compact code for the source is presented in Table 3.5.
The average codeword length is 1 bit, and the efficiency is

$$\eta_1 = 0.811.$$

Table 3.5   A compact code.

x_i	$p(x_i)$	Compact code
x_1	$\frac{3}{4}$	0
x_2	$\frac{1}{4}$	1

Example: To improve the efficiency, the second extension of the source is encoded, as shown in Table 3.6.

Table 3.6   A compact code for an extension of a source.

Symbol	Probability	Compact code
$x_1 x_1$	9/16	0
$x_1 x_2$	3/16	10
$x_2 x_1$	3/16	110
$x_2 x_2$	1/16	111

The average codeword length is $\frac{27}{16}$ bits. The extended source entropy is 2×0.811 bits, and the efficiency is

$$\eta_2 = \frac{2 \times 0.811 \times 16}{27} = 0.961.$$

The efficiency improves for each new extension of the original source, but, of course, the codes get longer, which implies that they take more time to transmit or process.

Example: The efficiencies associated with the third and fourth extensions of the source are

$$\eta_3 = 0.985$$

and

$$\eta_4 = 0.991.$$

As higher order extensions of the source are encoded, the efficiency approaches 1, a result that is proved in the next section.

3.5 Prefix Codes

For a prefix code, no codeword is a prefix, of the first part, of another codeword. Therefore, the code shown in Table 3.7 is a prefix. On the other hand, the code shown in Table 3.8 is not the prefix because the binary word 10, for instance, is a prefix for the codeword 100.

To decode a sequence of binary words produced by a prefix encoder, the decoder begins at the first binary digit of the sequence, and decodes a

Table 3.7 A prefix code for a given source.

Symbol	Code
x_1	1
x_2	01
x_3	001
x_4	000

Table 3.8 A source code that is not a prefix.

Symbol	Code
x_1	1
x_2	10
x_3	100
x_4	1000

codeword at a time. It is similar to a decision tree, which is a representation of the codewords of a given source code.

Figure 3.3 illustrates the decision tree for the prefix code pointed in Table 3.9.

The tree has one initial state and four final states, which correspond to the symbols x_1, x_2, and x_3. From the initial state, for each received bit, the decoder searches the tree until a final state is found.

The decoder then emits a corresponding decoded symbol and returns to the initial state. Therefore, from the initial state, after receiving a 1, the source decoder decodes symbol x_1 and returns to the initial state. If it receives a 0, the decoder moves to the lower part of the tree; in the following, after receiving another 0, the decoder moves further to the lower part of the tree and, after receiving a 1, the decoder retrieves x_2 and returns to the initial state.

Considering the code from Table 3.9, with the decoding tree from Figure 3.9, the binary sequence 0111000100101100101 is decoded into the output sequence $x_1 x_0 x_0 x_3 x_0 x_2 x_1 x_2 x_1$.

Table 3.9 Example of a prefix code.

Source symbol	Probability of occurrence	Code
x_0	0.5	1
x_1	0.25	01
x_2	0.125	001
x_3	0.125	000

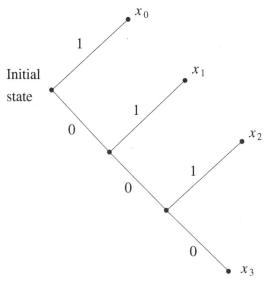

Figure 3.3 Decision tree for the code in Table 3.9.

By construction, a prefix code is always unequivocally decodable, which is important to avoid any confusion at the receiver.

Consider a code that has been constructed for a discrete source with alphabet $\{x_1, x_2, \ldots, x_K\}$. Let $\{p_1, p_1, \ldots, p_K\}$ be the source statistics and l_k be the codeword length for symbol x_k, $k = 1, \ldots, K$. If the binary code constructed for the source is a prefix one, then one can use the Kraft–McMillan inequality

$$\sum_{k=1}^{K} 2^{-l_k} \leq 1, \tag{3.34}$$

in which factor 2 is the radix, or number of symbols, of the binary alphabet.

For a memoryless discrete source with entropy $H(X)$, the codeword average length of a prefix code is limited to

$$H(X) \leq \overline{L} < H(X) + 1. \tag{3.35}$$

The left-hand side equality is obtained on the condition that symbol x_k be emitted from the source with probability $p_k = 2^{-l_k}$, in which l_k is the length of the codeword assigned to symbol x_k.

Consider the Nth order extension of a memoryless discrete source. For the N-th order extension of a code, the encoder operates on blocks of N

symbols from the original source, instead of individual ones, and the source alphabet X^N has an entropy that is N times the entropy of the original source.

Let $\overline{L}_N$ be the average length for the extended prefix code. For an unequivocally decodable code, $\overline{L}_N$ is as small as possible. From eqn 3.35, it follows that

$$H(X^N) \leq \overline{L}_N < H(X^N) + 1. \tag{3.36}$$

Therefore,

$$NH(X) \leq \overline{L}_N < NH(X) + 1 \tag{3.37}$$

or, in an equivalent way,

$$H(X) \leq \frac{\overline{L}_N}{N} < H(X) + \frac{1}{N}. \tag{3.38}$$

In the limit, as N goes to infinity, the inferior and superior limitants converge, and, therefore,

$$\lim_{N \to \infty} \frac{1}{N} \overline{L}_N = H(X). \tag{3.39}$$

Therefore, for a prefix extended encoder, as the order N increases, the code represents the source as efficiently as desired, and the average codeword length can be as close to the source entropy as possible, according to Shannon's source coding theorem. On the other hand, there is a compromise between the reduction on the average codeword length and the increase in complexity of the decoding process (Haykin, 1988).

3.6 The Information Unit

There is some confusion between the binary digit, abbreviated as bit, and the information particle, also baptized as bit by John Tukey and Claude Shannon.

In a meeting of the Institute of Electrical and Electronics Engineers (IEEE), the largest scientific institution in the world, the author of this book proposed the shannon [Sh] as a unit of information transmission, which is equivalent to bit per second. It is instructive to say that the bit, as used today, is not a unit of information because it is not approved by the International System of Units (SI).

What is curious about that meeting was the misunderstanding that surrounded the units, in particular, regarding the difference between the concepts of information unit and digital logic unit (Alencar, 2007).

To make things clear, the binary digit is associated with a certain state of a digital system, and not to information. A binary digit "1" can refer to 5 volts, in TTL logic, or 12 volts, for CMOS logic.

The information bit exists independent of any association with a particular voltage level. It can be associated, for example, with a discrete information or with the quantization of an analog information.

For instance, the information bits recorded on the surface of a compact disk are stored as a series of depressions on the plastic material, which are read by an optical beam, generated by a semiconductor laser. But, obviously, the depressions are not the information. They represent a means for the transmission of information, a material substrate that carries the data.

In the same way, the information can exist, even if it is not associated with light or other electromagnetic radiation. It can be transported by several means, including paper, and materializes itself when it is processed by a computer or by a human being.

4

Source Coding

> "You can't trust code that you did not totally create yourself."
>
> Ken Thompson

4.1 Types of Source Codes

This chapter presents the classification of source codes, block codes, non-singular codes, uniquely decodable codes, and instantaneouss codes.

4.1.1 Block Codes

Let $S = \{x_0, x_1, \ldots, x_{K-1}\}$ be a set of symbols of a given source alphabet. A code is defined as a mapping of all possible symbol sequences from S into sequences of another set $X = \{x_0, x_1, \ldots, x_{M-1}\}$, called the code alphabet.

A block code maps each of the symbols from the source set into a sequence of the code alphabet. The fixed sequences of symbols x_j are called codewords X_i. Note that X_i denotes a sequence of x_js (Abramson, 1963).

Example: A binary block code is presented in Table 4.1, and a ternary block code is shown in Table 4.2

4.1.2 Non-Singular Codes

A block code is said to be non-singular if all codewords are distinct (Abramson, 1963). Table 4.2 shows an example of a non-singular code. The code shown in Table 4.3 is also non-singular, but although the

Table 4.1   A binary block code.

Source symbols	Code
x_0	0
x_1	11
x_2	00
x_3	1

Table 4.2   A ternary block code.

Source symbols	Code
x_0	0
x_1	1
x_2	2
x_3	01

Table 4.3   A non-singular block code.

Source symbols	Code
x_0	0
x_1	01
x_2	1
x_3	11

codewords are distinct, there is a certain ambiguity between some symbol sequences of the code regarding the source symbol sequences.

Example: The sequence 1111 can correspond to $x_2x_2x_2x_2$, $x_2x_3x_2$, or even x_3x_3, which indicates that it is necessary to define a more strict condition than non-singularity for a code, to guarantee that it can be used in a practical situation.

4.1.3 Uniquely Decodable Codes

Let a block code map symbols from a source alphabet S into fixed symbol sequences of a code alphabet X. The source can be an extension of another source, which is composed of symbols from the original alphabet. The n-ary extension of a block code that maps symbols x_i into codewords X_i is the

Table 4.4 A non-singular block code.

Source symbols	Code
x_0	1
x_1	00
x_2	11
x_3	10

Table 4.5 The second extension of a block code.

Source symbols	Code	Source symbols	Code
x_0x_0	11	x_2x_0	111
x_0x_1	100	x_2x_1	1100
x_0x_2	111	x_2x_2	1111
x_0x_3	110	x_2x_3	1110
x_1x_0	001	x_3x_0	101
x_1x_1	0000	x_3x_1	1000
x_1x_2	0011	x_3x_2	1011
x_1x_3	0010	x_3x_3	1010

block code that maps symbol sequences from the source $(x_{i1}x_{i2} \ldots x_{in})$ into the codeword sequences $(X_{i1}X_{i2} \ldots X_{in})$ (Abramson, 1963).

From the previous definition, the n-ary extension of a block code is also a block code. The second-order extension of the block code presented in Table 4.4 is the block code of Table 4.5.

A block code is said to be uniquely decodable if and only if the n-ary extension of the code is non-singular for all finite n.

4.1.4 Instantaneous Codes

Table 4.6 presents two examples of uniquely decodable codes. Code $\mathcal{A}$ is a simpler method to construct a uniquely decodable set of sequences because all codewords have the same length and it is a non-singular code.

Code $\mathcal{B}$ is also uniquely decodable. It is also called a comma code because the digit zero is used to separate the codewords (Abramson, 1963).

Consider the code shown in Table 4.7. Code $\mathcal{C}$ differs from $\mathcal{A}$ and $\mathcal{B}$ from Table 4.6 in an important aspect. If a binary sequence composed of codewords from code $\mathcal{C}$ occurs, it is not possible to decode the sequence.

Table 4.6 Uniquely decodable codes.

Source symbols	Code $\mathcal{A}$	Code $\mathcal{B}$
x_0	000	0
x_1	001	10
x_2	010	110
x_3	011	1110
x_4	100	11110
x_5	101	111110
x_6	110	1111110
x_7	111	11111110

Table 4.7 Another uniquely decodable code.

Source symbols	Code $\mathcal{C}$
x_0	1
x_1	10
x_2	100
x_3	1000
x_4	10000
x_5	100000
x_6	1000000
x_7	10000000

Example: If the bit stream 100000 is received, for example, it is not possible to decide if it corresponds to symbol x_5, unless the next symbol is available. If the next symbol is 1, then the sequence is 100000, but if it is 0, then it is necessary to inspect one more symbol to know if the sequence corresponds to x_6 (1000000) or x_7 (10000000).

A uniquely decodable code is instantaneous if it is possible to decode each codeword in a sequence with no reference to subsequent symbols (Abramson, 1963). Codes $\mathcal{A}$ and $\mathcal{B}$ are instantaneous, and $\mathcal{C}$ is not.

It is possible to devise a test to indicate when a code is instantaneous. Let $X_i = x_{i1}x_{i2} \ldots x_{im}$ be a word from a certain code. The sequence of symbols $(x_{i1}x_{i2} \ldots x_{ij})$, with $j \leq m$, is called the prefix of the codeword X_i.

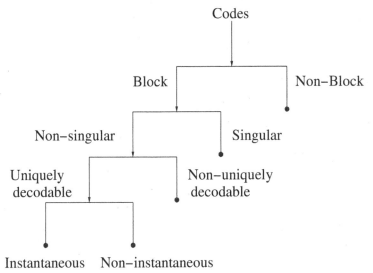

Figure 4.1 Classes of source codes.

Example: The codeword 10000 has five prefixes: 1, 10, 100, 1000, and 10000. A necessary condition for a code to be instantaneous is that no codeword is a prefix of another codeword.

The various classes of codes presented in this section are summarized in Figure 4.1.

4.2 Construction of Instantaneous Codes

In order to construct a binary instantaneous code for a source with five symbols, one can begin by attributing the digit 0 to symbol s_0 (Abramson, 1963)

$$x_0 \leftarrow 0.$$

In this case, the remaining source symbols should correspond to the codewords that begin with the digit 1. Otherwise, it is not a prefix code. It is not possible to associate x_1 with the codeword 1 because no other symbol would remain to begin the other codewords.

Therefore,

$$x_1 \leftarrow 10.$$

This codeword assignment requires that the remaining codewords begin with 11. If

$$x_2 \leftarrow 110,$$

then, the only unused prefix with 3 bits is 111, which implies that

$$x_3 \leftarrow 1110$$

and

$$x_4 \leftarrow 1111.$$

In the previously constructed code, note that if one begins the code construction by making x_0 to correspond to 0, this restricts the available number of codewords, because the remaining codewords had to, necessarily, begin with 1.

On the other hand, if a two-digit word had been chosen to represent x_0, there would be more freedom to choose the others, and there would be no need to assign very long codewords to the last ones.

A binary instantaneous code can be constructed to represent the five symbols (Abramson, 1963). The first assignment is

$$x_0 \leftarrow 00.$$

Then, one can assign

$$x_1 \leftarrow 01$$

and two unused prefixes of length 2 are saved to the following codeword assignment:

$$x_2 \leftarrow 10$$

$$x_3 \leftarrow 110$$

$$x_4 \leftarrow 111.$$

The question of which code is the best is postponed for the next section, because it requires the notion of average length of a code, which depends on the symbol probability distribution.

4.3 Kraft Inequality

Consider an instantaneous code with source alphabet given by

$$S = \{x_0, x_1, \ldots, x_{K-1}\}$$

and code alphabet

$$X = \{x_0, x_1, \ldots, x_{M-1}\}.$$

Let $X_0, X_1, \ldots, X_{K-1}$ be the codewords, and let l_i be the length of the word X_i. The Kraft inequality establishes that a necessary and sufficient condition for the existence of an instantaneous code of length $l_0, l_1, \ldots, l_{K-1}$ is

$$\sum_{i=0}^{K-1} r^{-l_i} \leq 1, \tag{4.1}$$

in which r is the number of different symbols of the code.

For the binary case,

$$\sum_{i=0}^{K-1} 2^{-l_i} \leq 1. \tag{4.2}$$

The Kraft inequality can be used to determine if a given sequence of length l_i is acceptable for a codeword of an instantaneous code.

Consider an information source, with four possible symbols, x_0, x_1, x_2, and x_3. Table 4.8 presents five possible codes to represent the original symbols, using a binary alphabet.

Example: For code $\mathcal{A}$, one obtains

$$\sum_{i=0}^{3} 2^{-l_i} = 2^{-2} + 2^{-2} + 2^{-2} + 2^{-2} = 1.$$

Therefore, the codeword lengths of this code are acceptable for an instantaneous code. But, the Kraft inequality does not tell if $\mathcal{A}$ is an instantaneous code. It is only a necessary condition that has to be fulfilled by the lengths.

For example, the inequality states that there is an instantaneous code with four codewords of length 2. In this case, it is clear that the binary codewords of code $\mathcal{A}$ satisfy the Kraft inequality and also form an instantaneous code.

Table 4.8   Selected binary codes.

Source symbols	Code $\mathcal{A}$	Code $\mathcal{B}$	Code $\mathcal{C}$	Code $\mathcal{D}$	Code $\mathcal{E}$
x_0	11	1	1	1	1
x_1	10	011	01	011	01
x_2	01	001	001	001	001
x_3	00	000	000	00	00

For code $\mathcal{B}$,

$$\sum_{i=0}^{3} 2^{-l_i} = 2^{-1} + 2^{-3} + 2^{-3} + 2^{-3} = 7/8 \le 1.$$

In this case, the lengths of the codewords are suitable to compose an instantaneous code. Code $\mathcal{B}$ is also a prefix code.

Code $\mathcal{C}$ is similar to code $\mathcal{B}$, except for a discarded bit in the second codeword. For this code, one obtains

$$\sum_{i=0}^{3} 2^{-l_i} = 2^{-1} + 2^{-2} + 2^{-3} + 2^{-3} = 1.$$

The codeword lengths satisfy the Kraft inequality and, by inspection, one observes that this code is instantaneous.

Code $\mathcal{D}$ is obtained from $\mathcal{B}$, discarding a bit in the fourth codeword. Although the lengths satisfy the Kraft inequality, code $\mathcal{D}$ is not instantaneous because it is not a prefix code. The fourth codeword is a prefix of the third one.

Finally, for code $\mathcal{E}$,

$$\sum_{i=0}^{3} 2^{-l_i} = \frac{9}{8},$$

and the codeword lengths do not satisfy the Kraft inequality. Therefore, code $\mathcal{E}$ is not instantaneous.

Consider a source with eight symbols to be encoded into an instantaneous ternary code, whose codeword lengths are $1, 2, 2, 2, 2, 3, 3, 3$. Using the Kraft inequality,

$$\sum_{i=0}^{9} 3^{-l_i} = \frac{1}{3} + 4\frac{1}{9} + 3\frac{1}{27} = \frac{24}{27} < 1,$$

which indicates that this code is possible, as follows:

$$x_0 \leftarrow 0$$

$$x_1 \leftarrow 10$$

$$x_2 \leftarrow 11$$

$$x_3 \leftarrow 20$$

$$x_4 \leftarrow 21$$

$$x_5 \leftarrow 220$$

$$x_6 \leftarrow 221$$

$$x_7 \leftarrow 222.$$

For a source with 11 symbols, if the codeword lengths are 1, 2, 2, 2, 2, 2, 2, 3, 3, 3, 3, it is not possible to obtain a ternary instantaneous code because

$$\sum_{i=0}^{10} 3^{-l_i} = \frac{1}{3} + 6\frac{1}{9} + 4\frac{1}{27} = \frac{31}{27} > 1.$$

4.4 Huffman Code

This section describes the Huffman coding algorithm, and the procedure to construct the Huffman code when the source statistics are known.

The technique was developed by David Albert Huffman (1925–1999), in a paper for a course on information theory taught by Robert Mario Fano (1917-), at the Massachusetts Institute of Technology (MIT). The obtained sequences are called Huffman codes, and they are prefix codes.

Huffman procedure is based on two assumptions regarding the optimum prefix codes:

1. The most frequent symbols, those with higher probability, are represented by shorter codewords.
2. The least frequent symbols are assigned codewords of same length.

According to the first assumption, as the most probable symbols are also the most frequent, they must be as short as possible to decrease the average length of the code. The second assumption is also true because for a prefix code, a shorter codeword could not be a prefix of another one. The least probable symbols must be distinct and have same length (Sayood, 2006).

Furthermore, the Huffman process is completed by the addition of a simple requisite. The longer codewords that correspond to the least frequent symbols differ only on the last digit.

4.4.1 Constructing a Binary Huffman Code

Given a discrete source, a Huffman code can be constructed along the following steps:

1. The source symbols are arranged in decreasing probability. The least probable symbols receive the assignments 0 and 1.
2. Both symbols are combined to create a new source symbol, whose probability is the sum of the original ones. The list is reduced by one symbol. The new symbol is positioned in the list according to its probability.
3. This procedure continues until the list has only two symbols, which receive the assignments 0 and 1.
4. Finally, the binary codeword for each symbol is obtained by a reverse process.

In order to explain the algorithm, consider the source of Table 4.9.

The first phase is to arrange the symbols in a decreasing order of probability. Assign the values 0 and 1 to the symbols with the smallest probabilities. They are, then, combined to create a new symbol. The probability associated with the new symbol is the sum of the previous probabilities. The new symbol is repositioned in the list, to maintain the same decreasing order for the probabilities. The procedure is shown in Figure 4.2.

The procedure is repeated until only two symbols remain, which are assigned to 0 and 1, as shown in Figure 4.3.

The procedure is repeated to obtain all codewords, by reading the digits in inverse order, from Phase IV to Phase I, as illustrated in Figure 4.4. Following the arrows, for symbol x_4, one finds the codeword 011.

Table 4.9   Discrete source with five symbols and their probabilities.

Symbols	Probabilities
x_0	0.4
x_1	0.2
x_2	0.2
x_3	0.1
x_4	0.1

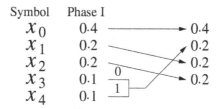

Figure 4.2 Probabilities in descending order for the Huffman code.

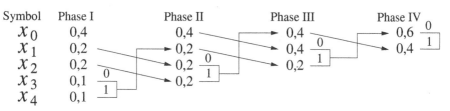

Figure 4.3 Huffman code. At each phase, the two least probable symbols are combined.

For the example, the average codeword length for the Huffman code is given by

$$\overline{L} = \sum_{i=0}^{4} p_k l_k = 0.4(2) + 0.2(2) + 0.2(2) + 0.1(3) + 0.1(3) = 2.2 \text{ bits.}$$

The source entropy is calculated as

$$H(X) = \sum_{i=0}^{4} p_k \log_2 \left(\frac{1}{p_k} \right)$$

$$H(X) = 0.4 \log_2 \left(\frac{1}{0.4} \right) + 0.2 \log_2 \left(\frac{1}{0.2} \right) + 0.2 \log_2 \left(\frac{1}{0.2} \right)$$

$$+ \ 0.1 \log_2 \left(\frac{1}{0.1} \right) + 0.1 \log_2 \left(\frac{1}{0.1} \right)$$

or

$$H(X) = 2.12193 \text{ bits.}$$

The code efficiency is

$$\eta = \frac{H(X)}{\overline{L}} = \frac{2.12193}{2.2},$$

which is equal to 96.45%.

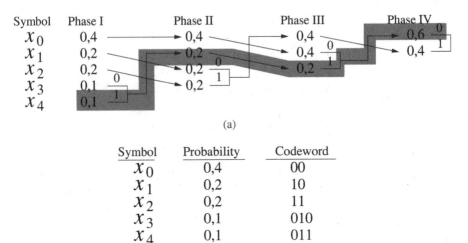

(a)

Symbol	Probability	Codeword
x_0	0,4	00
x_1	0,2	10
x_2	0,2	11
x_3	0,1	010
x_4	0,1	011

(b)

Figure 4.4   (a) Example of the Huffman coding algorithm to obtain the codewords. (b) Resulting code.

It is important to say that the Huffman procedure is not unique, and several variations can be obtained for the final set of codewords, depending on the way the bits are assigned. But, in spite of how the probabilities are positioned, the average length is always the same, if the rules are followed.

The difference is the variance of the codeword lengths, defined as

$$V[L] = \sum_{k=0}^{K-1} p_k (l_k - \overline{L})^2, \tag{4.3}$$

in which p_k and l_k denote the probability of occurrence of the kth source symbol and length of the respective codeword.

Usually, the procedure of displacing the probability of the new symbol to the highest position in the list produces smaller values for $V[L]$, as compared to the displacement of the probability to the lowest position of the list.

Table 4.10 presents four Huffman codes obtained for the source of Table 4.9. Codes I and II were obtained by shifting the new symbol to the highest position in the list of decreasing probabilities.

Codes III and IV were produced by shifting the new symbol to the lowest position in the list. Codes I and III used the systematic assignment of 0 followed by 1 to the least frequent symbols. Codes II and IV used the systematic assignment of 1 followed by 0 to the least frequent symbols. For all codes, the average codeword length is 2.2 bits. For codes I and II, the variance of the codeword lengths is 0.16. For codes III and IV, the variance is 1.36.

Table 4.10 Four distinct Huffman codes obtained for the source of Table 4.9.

Symbols	Code I	Code II	Code III	Code IV
x_0	00	11	1	0
x_1	10	01	01	10
x_2	11	00	000	111
x_3	010	101	0010	1101
x_4	011	100	0011	1100

5

Information Transmission and Channel Capacity

"One of the most singular characteristics of the art of deciphering is the strong conviction possessed by every person, even moderately acquainted with it, that he is able to construct a cipher which nobody else can decipher. I have also observed that the cleverer the person, the more intimate is his conviction."

Charles Babbage, Passages from the Life of a Philosopher

Claude Elwood Shannon (1916–2001) is considered the father of information theory. In 1948, he published a seminal article on the mathematical concept of information, which is one of the most cited for decades. Information left the Journalism field to occupy a more formal area, as part of probability theory.

The entropy, in the context of information theory, was initially defined by Ralph Vinton Lyon Hartley (1888–1970), in the article "Transmission of Information," published by the Bell System Technical Journal, in July 1928, ten years before the formalization of the concept by Claude Shannon.

Shannon's development was also based on Harry Nyquist's work (Harry Theodor Nyquist, 1889–1976), which determined the sampling rate as a function of frequency necessary to reconstruct an analog signal using a set of discrete samples.

In an independent way, Andrei N. Kolmogorov developed his complexity theory, during the 1960 decade. It was a new information theory based on the length of an algorithm developed to describe a certain data sequence. He used Alan Turing's machine in this new definition. Under certain conditions, Kolmogorov's and Shannon's definitions are equivalent.

The idea of relating the number of states of a system with a physical measure, although, dates back to the XIX century. Rudolph Clausius proposed the term entropy for such a measure in 1895.

Entropy comes from the Greek word for transformation and, in physics, is related to the logarithm of the ratio between the final and initial temperature of a system or to the ratio of the heat variation and the temperature of the same system.

Shannon defined the entropy of an alphabet at the negative of the mean value of the logarithm of the symbols' probability. This way, when the symbols are equiprobable, the definition is equivalent to Nyquist's.

But, as a more generic definition, Shannon's entropy can be used to compute the capacity of communication channels. Most part the researchers' work is devoted to either compute the capacity or to develop error correcting codes to attain that capacity.

Shannon died on February 24, 2001, as a victim of a disease named after the physician Aloysius Alzheimer. According to his wife, he lived a quiet life but had lost his capacity to retain information.

5.1 The Concept of Information Theory

The concept of information transmission is associated with the existence of a communication channel that links the source and destination of the message. This can imply the occurrence of transmission errors, caused by the probabilistic nature of the channel.

Figure 5.1 illustrates the canonical model for a communication channel, proposed by Shannon in his seminal paper of 1948 Shannon, 1948b. This is a very simplified model of reality but contains the basic blocks upon which the mathematical structure is built.

5.2 Joint Information Measurement

Consider two discrete and finite sample spaces, Ω and Ψ, with the associated random variables X and Y,

$$
\begin{aligned}
X &= x_1, x_2, \ldots, x_N, \\
Y &= y_1, y_2, \ldots, y_M.
\end{aligned}
\tag{5.1}
$$

The events from Ω may jointly occur with events from Ψ. Therefore, the following matrix contains the whole set of events in the product space $\Omega\Psi$,

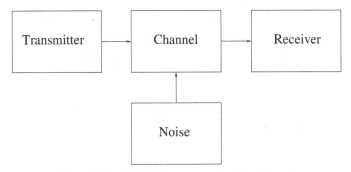

Figure 5.1   Model for a communication channel.

$$
[XY] = \begin{bmatrix} x_1y_1 & x_1y_2 & \cdots & x_1y_M \\ x_2y_1 & x_2y_2 & \cdots & x_2y_M \\ \cdots & \cdots & \cdots & \cdots \\ x_Ny_1 & x_Ny_2 & \cdots & x_Ny_M \end{bmatrix} \tag{5.2}
$$

The joint probability matrix is given in the following, in which no restriction is assumed regarding the dependence between the random variables

$$
[P(X,Y)] = \begin{bmatrix} p_{1,1} & p_{1,2} & \cdots & p_{1,M} \\ p_{2,1} & p_{2,2} & \cdots & p_{2,M} \\ \cdots & \cdots & \cdots & \cdots \\ p_{N,1} & p_{N,2} & \cdots & p_{N,M} \end{bmatrix} \tag{5.3}
$$

Figure 5.2 shows the relation between the input and output alphabets, which are connected by the joint probability distribution matrix $[P(X,Y)]$.

The joint entropy between the random variables from sources X and Y is given by

$$
H(X,Y) = - \sum_{k=1}^{N} \sum_{j=1}^{M} p_{k,j} \log p_{k,j}, \tag{5.4}
$$

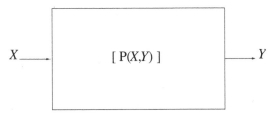

Figure 5.2   A probabilistic communication channel.

which may be simplified to

$$H(X,Y) = -\sum_X \sum_Y p(x,y) \log p(x,y). \qquad (5.5)$$

The marginal entropies may be written in terms of the marginal probabilities, $p(x)$ e $p(y)$

$$H(X) = -\sum_X p(x) \log p(x) \qquad (5.6)$$

and

$$H(Y) = -\sum_Y p(y) \log p(y). \qquad (5.7)$$

5.3 Conditional Entropy

The concept of conditional entropy is essential to model, and understand, the operation of the communication channel because it provides information about a particular symbol, given that another symbol has occurred. The entropy of alphabet X conditioned to the occurrence of a particular symbol y is given by

$$H(X|y) = -\sum_X \frac{p(x,y)}{p(y)} \log \frac{p(x,y)}{p(y)}$$

$$= -\sum_X p(x|y) \log p(x|y). \qquad (5.8)$$

The expected value of the conditional entropy, for all possibles values of y, provides the average conditional entropy of the system

$$H(X|Y) = E[H(X|y)] = \sum_Y p(y)\,[H(X|y)]$$

$$= -\sum_Y p(y) \sum_X p(x|y) \log p(x|y), \qquad (5.9)$$

which can be written as

$$H(X|Y) = -\sum_Y \sum_X p(y)p(x|y) \log p(x|y), \qquad (5.10)$$

or

$$H(X|Y) = - \sum_Y \sum_X p(x,y) \log p(x|y). \qquad (5.11)$$

In the same way, the mean conditional entropy of source Y, given the information about source X, is

$$H(Y|X) = - \sum_X \sum_Y p(x)p(y|x) \log p(y|x) \qquad (5.12)$$

or

$$H(Y|X) = - \sum_X \sum_Y p(x,y) \log p(y|x). \qquad (5.13)$$

5.4 Model for a Communication Channel

A communication channel can be modeled based on the previous developments. Consider a source that has the given alphabet X. The source transmits the information to the destiny using a certain channel. The system may be described by a joint probability matrix, which gives the joint probability of occurrence of a transmitted symbol and a received one,

$$[P(X,Y)] = \begin{bmatrix} p(x_1,y_1) & p(x_1,y_2) & \cdots & p(x_1,y_N) \\ p(x_2,y_1) & p(x_2,y_2) & \cdots & p(x_2,y_N) \\ \cdots & \cdots & \cdots & \cdots \\ p(x_M,y_1) & p(x_M,y_2) & \cdots & p(x_M,y_N) \end{bmatrix} \qquad (5.14)$$

There are five probability schemes to analyze:

1. $[P(X,Y)]$, joint probability matrix;
2. $[P(X)]$, marginal probability matrix of X;
3. $[P(Y)]$, marginal probability matrix of Y;
4. $[P(X|Y)]$, probability matrix conditioned on Y;
5. $[P(Y|X)]$, probability matrix conditioned on X;

Those probability schemes produce five entropy functions, associated with the communication channel, whose interpretations are given as follows:

1. $H(X)$ – Average information per source symbol, or source entropy;
2. $H(Y)$ – Average information per received symbol, or receiver entropy;
3. $H(X,Y)$ – Average information associated with pairs of transmitted and received symbols, or average uncertainty of the communication system;
4. $H(X|Y)$ – Average information measurement of the received symbol, given that X was transmitted, or conditional entropy;

5. $H(Y|X)$ – Average information measurement of the source, given that Y was received, or equivocation.

5.5 Noiseless Channel

For the noiseless discrete channel, each symbol from the input alphabet has a one-to-one correspondence with the output. The joint probability matrix as well as the transition probability matrix have the same diagonal format

$$[P(X,Y)] = \begin{bmatrix} p(x_1,y_1) & 0 & \cdots & 0 \\ 0 & p(x_2,y_2) & \cdots & 0 \\ \cdots & \cdots & \cdots & \cdots \\ 0 & 0 & \cdots & p(x_N,y_N) \end{bmatrix} \quad (5.15)$$

$$[P(X|Y)] = [P(Y|X)] = \begin{bmatrix} 1 & 0 & \cdots & 0 \\ 0 & 1 & \cdots & 0 \\ \cdots & \cdots & \cdots & \cdots \\ 0 & 0 & \cdots & 1 \end{bmatrix} \quad (5.16)$$

The joint entropy equals the marginal entropies

$$H(X,Y) = H(X) = H(Y) = -\sum_{i=1}^{N} p(x_i,y_i) \log p(x_i,y_i), \quad (5.17)$$

and the conditional entropies are null

$$H(Y|X) = H(X|Y) = 0. \quad (5.18)$$

As a consequence, the receiver uncertainty is equal to the source entropy, and there is no ambiguity at the reception, which indicates that the conditional entropies are all zero.

5.6 Channel with Independent Output and Input

For the channel with independent input and output there is no relation between the transmitted and received symbols, that is, given that a given symbol has been transmitted, any symbol can be received, with no connection whatsoever with it. The joint probability matrix has N identical columns

$$[P(X,Y)] = \begin{bmatrix} p & p_1 & \cdots & p_1 \\ p_2 & p_2 & \cdots & p_2 \\ \cdots & \cdots & \cdots & \cdots \\ p_M & p_M & \cdots & p_M \end{bmatrix}, \qquad \sum_i^M p_i = \frac{1}{N}. \qquad (5.19)$$

The input and output symbol probabilities are statistically independent, that is,

$$p(x,y) = p(x)p(y). \qquad (5.20)$$

Computing the entropy gives

$$H(X,Y) = -N \left(\sum_{i=1}^{M} p_i \log p_i \right), \qquad (5.21)$$

$$H(X) = -\sum_{i=1}^{M} Np_i \log Np_i = -N \left(\sum_{i=1}^{M} p_i \log p_i \right) - \log N, \qquad (5.22)$$

$$H(Y) = -N \left(\frac{1}{N} \log \frac{1}{N} \right) = \log N, \qquad (5.23)$$

$$H(X|Y) = -\sum_{i=1}^{M} Np_i \log Np_i = H(X), \qquad (5.24)$$

$$H(Y|X) = -\sum_{i=1}^{M} Np_i \log \frac{1}{N} = \log N = H(Y). \qquad (5.25)$$

As a consequence, the channel with independent input and output does not provide information, that is, it has the highest possible loss, contrasting with the noiseless channel.

5.7 Relations Between the Entropies

It is possible to show, using Bayes rule for the conditional probability, that the joint entropy can be written in terms of the conditional entropy, in the following way:

$$H(X,Y) = H(X|Y) + H(Y), \qquad (5.26)$$
$$H(X,Y) = H(Y|X) + H(X). \qquad (5.27)$$

Shannon has shown the fundamental inequality

$$H(X) \geq H(X|Y), \tag{5.28}$$

whose demonstration is given in the following.

The logarithm concavity property can be used to demonstrate the inequality, $\ln x \leq x - 1$, as follows:

$$H(X|Y) - H(X) = \sum_Y \sum_X p(x,y) \log \frac{p(x)}{p(x|y)}$$

$$\leq \sum_Y \sum_X p(x,y) \left(\frac{p(x)}{p(x|y)} - 1 \right) \log e. \tag{5.29}$$

But, the right-hand side of the inequality is zero, as shown in the following:

$$\sum_Y \sum_X (p(x) \cdot p(y) - p(x,y)) \log e = \sum_Y (p(y) - p(y)) \log e$$

$$= 0. \tag{5.30}$$

Therefore,

$$H(X) \geq H(X|Y). \tag{5.31}$$

In a similar manner, it can be shown that

$$H(Y) \geq H(Y|X). \tag{5.32}$$

The equality is attained if and only if X and Y are statistically independent.

5.8 Mutual Information

A measure of mutual information provided by two symbols (x_i, y_i) can be written as

$$I(x_i; y_j) = \log_2 p(x_i|y_j) - \log_2 p(x_i)$$

$$= \log_2 \frac{p(x_i|y_j)}{p(x_i)} = \log \frac{p(x_i, y_j)}{p(x_i)p(y_j)}. \tag{5.33}$$

It can be noticed that the *a priori* information of symbol x_i is contained in the marginal probability $p(x_i)$. The *a posteriori* probability that symbol

x_i has been transmitted, given that y_i was received, is $p(x_i|y_i)$. Therefore, in an informal way, the information gain for the observed symbol equals the difference between the initial information, or uncertainty, and the final one.

The mutual information is continuous in $p(x_i|y_i)$, and also symmetric, or

$$I(x_i; y_j) = I(y_j; x_i), \tag{5.34}$$

which indicates that the information provided by x_i about y_i is the same provided by y_i about x_i.

The function $I(x_i; x_i)$ can be called the auto-information of x_i, or

$$I(x_i) = I(x_i; x_i) = \log \frac{1}{p(x_i)} \tag{5.35}$$

because, for an observer of the source alphabet, the *a priori* knowledge of the situation is that x_i will be transmitted with probability $p(x_i)$, and the *a posteriori* knowledge is the certainty that x_i transmitted.

In conclusion,

$$I(x_i; y_j) \leq I(x_i; x_i) = I(x_i), \tag{5.36}$$
$$I(x_i; y_j) \leq I(y_j; y_j) = I(y_j). \tag{5.37}$$

The statistical mean of the mutual information per pairs of symbols provides an interesting interpretation of the mutual information concept

$$I(X; Y) = E[I(x_i; y_j)] = \sum_i \sum_j p(x_i, y_j) I(x_i; y_j), \tag{5.38}$$

which can be written as

$$I(X; Y) = \sum_i \sum_j p(x_i, y_j) \log \frac{p(x_i|y_j)}{p(x_i)}. \tag{5.39}$$

The average mutual information can be interpreted as a reduction on the uncertainty about the input X, when the output Y is observed (MacKay, 2003). This definition provides an adequate metric for the average mutual information of all pairs of symbols and can be put in terms of the entropy, such as

$$I(X; Y) = H(X) + H(Y) - H(X,Y), \tag{5.40}$$
$$I(X; Y) = H(X) - H(X|Y), \tag{5.41}$$
$$I(X; Y) = H(Y) - H(Y|X). \tag{5.42}$$

Put that way, the average mutual information gives a measure of the information that is transmitted by the channel. Because of this, it is called transinformation, or information transferred by the channel. It is always non-negative, even if the individual information quantities are negative for certain pairs of symbols.

For a noiseless channel, the average mutual information equals the joint entropy.

$$I(X;Y) = H(X) = H(Y), \tag{5.43}$$
$$I(X;Y) = H(X,Y). \tag{5.44}$$

On the other hand, for a channel in which the output is independent of the input, the average mutual information is null, implying that no information is transmitted by the channel

$$I(X;Y) = H(X) - H(X|Y),$$
$$= H(X) - H(X) = 0. \tag{5.45}$$

It is possible to obtain relations between the entropies of a multiple-port channel. For a three-port channel, for example, one can obtain the following relations for the entropies:

$$H(X,Y,Z) \leq H(X) + H(Y) + H(Z), \tag{5.46}$$
$$H(Z|X,Y) \leq H(Z|Y). \tag{5.47}$$

In the same reasoning, it is possible to obtain the following relations for the average mutual information for a three-port channel:

$$I(X;Y,Z) = I(X;Y) + I(X;Z|Y), \tag{5.48}$$
$$I(Y,Z;X) = I(Y;X) + I(Z;X|Y). \tag{5.49}$$

5.9 Channel Capacity

Shannon defined the discrete channel capacity as the maximum of the average mutual information, computed for all possible probability sets that can be associated with the input symbol alphabet, that is, for all memoryless sources,

$$C = \max I(X;Y) = \max [H(X) - H(X|Y)]. \tag{5.50}$$

5.9.1 Capacity of the Memoryless Discrete Channel

Consider X as the alphabet of a source with N symbols. Because the transition probability matrix is diagonal, one obtains

$$C = \max I(X;Y) = \max[H(X)] = \max\left[-\sum_{i=1}^{N} p(x_i)\log p(x_i)\right]. \quad (5.51)$$

Example: The entropy attains a maximum when all symbols are equiprobable. Therefore, for the memoryless discrete channel, the capacity is

$$C = \max\left[-\sum_{i=1}^{N} \frac{1}{N}\log\frac{1}{N}\right],$$

which gives

$$C = \log N \text{ bits per symbol.} \quad (5.52)$$

The channel capacity can also be expressed in bits per second, or shannon (Sh), and corresponds to the information transmission rate of the channel, for symbols with duration of T seconds,

$$C_T = \frac{C}{T} \text{bits per second, or Sh.} \quad (5.53)$$

Therefore, for the noiseless channel,

$$C_T = \frac{C}{T} = \frac{1}{T}\log N \text{bits per second, or Sh.} \quad (5.54)$$

5.9.2 Relative Redundancy and Efficiency

The absolute redundancy is the difference between the actual information transmission rate and $I(X;Y)$ and the maximum possible value,

$$\text{Absolute redundancy for a noisy channel} = C - I(X;Y)$$
$$= \log N - H(X). \quad (5.55)$$

The ratio between the absolute redundancy and the channel capacity is defined as the system relative redundancy,

Relative redundancy for a noiseless channel, $D = \dfrac{\log N - H(X)}{\log N}$

$$= 1 - \frac{H(X)}{\log N}. \quad (5.56)$$

The system efficiency is defined as the complement of the relative redundancy,

Efficiency of the noiseless channel, $E = \dfrac{I(X;Y)}{\log N} = \dfrac{H(X)}{\log N}$

$$= 1 - D. \quad (5.57)$$

When the transmitted symbols do not occupy the same time interval, it is still possible to define the average information transmission rate for the noiseless channel as

$$R_T = \frac{-\displaystyle\sum_{i=1}^{N} p(x_i)\log p(x_i)}{\displaystyle\sum_{i=1}^{N} p(x_i)T_i}, \quad (5.58)$$

in which T_i represents the symbol intervals.

For a discrete noisy channel, the capacity is the maximum of the average mutual information, when the noise characteristic $p(y_i|x_i)$ is specified

$$C = \max \left(\sum_{i=1}^{N} \sum_{j=1}^{M} p(x_i)\, p(y_j|x_i) \log \frac{p(y_j|x_i)}{p(y_j)} \right), \quad (5.59)$$

in which the maximum is over $p(x_i)$. It must be noticed that the maximization in relation to the input probabilities do not always lead to an admissible set of source probabilities.

Bayes' rule defines the relation between the marginal probabilities $p(y_j)$ and the *a priori* probabilities $p(x_i)$,

$$p(y_j) = \sum_{i=1}^{N} p_1(x_i)\, p(y_j|x_i), \quad (5.60)$$

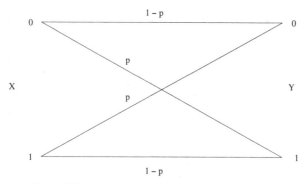

Figure 5.3 Memoryless binary symmetric channel.

in which the variables are restricted to the following conditions:

$$p(x_i) \geq 0 \qquad i = 1, 2, \ldots, N, \qquad (5.61)$$

$$\sum_{i=1}^{N} p_1(x_i) = 1.$$

Example: Determine the capacity of the memoryless binary symmetric channel (BSC) (Blake, 1987).

The channel is illustrated in Figure 5.3, in which the error probability, or the probability of transmitting a symbol and receiving another one, is indicated as p.

The channel capacity is given by Formula(5.50),

$$C = \max I(X;Y),$$

in which the average mutual information can be written as

$$I(X;Y) = \sum_{i=0}^{1} \sum_{j=0}^{1} p_{ij} \log \frac{p_{ij}}{p_i q_j}. \qquad (5.62)$$

Assume that the symbol *a priori* probabilities are $p_0 = r$ and $p_1 = v$, with $r + v = 1$. Probabilities r and v are chosen in such a manner that, for each channel use, the maximum quantity of information is transferred.

The joint probabilities are

$$p_{00} = r(1-p), \ p_{01} = rp, \ p_{10} = (1-r)p, \ p_{11} = (1-r)(1-p).$$

The average mutual information can be written as

$$
\begin{aligned}
I(X;Y) = \; & [(1-p)r] \log \left(\frac{(1-p)r}{r[(1-p)r + (1-r)p]} \right) \\
& + [rp] \log \left(\frac{rp}{r[(rp + (1-r)(1-p)]} \right) \\
& + [(1-r)p] \log \left(\frac{(1-r)p}{(1-r)[r(1-p) + (1-r)p]} \right) \\
& + [(1-p)(1-r)] \log \left(\frac{(1-p)(1-r)}{(1-r)[(rp + (1-r)(1-p)]} \right),
\end{aligned}
$$

which can be put in the following way, after the simplification with logarithm properties:

$$
\begin{aligned}
I(X;Y) = \; & p \log p + (1-p) \log (1-p) \\
& - [r(1-p) + (1-r)p] \log [r(1-p) + (1-r)p] \\
& + [rp + (1-r)(1-p)] \log [rp + (1-r)(1-p)].
\end{aligned}
$$

The objective is to determine the value of r that maximizes the expression, taking into account that the logarithms are base two. The obtained expression for r is a complicated one, but the maximum that the average mutual information attains is given by

$$
C = \max I(X;Y) = 1 - p \log p + (1-p) \log (1-p), \qquad (5.63)
$$

which represents the memoryless binary symmetric channel capacity. The graph for the capacity $C(p)$, as a function of the channel error probability, is shown in Figure 5.4.

Example: Determine the capacity of the binary erasure channel (BEC) shown in Figure 5.5, in which the parameter E represents the erasure and $1-p$ the occurrence probability (Blake, 1987).

The average mutual information for this channel is given by

$$
I(X;Y) = \sum_{i=0}^{1} \sum_{j=0}^{2} p_{ij} \log \frac{p_{ij}}{p_i q_j}, \qquad (5.64)
$$

and the probabilities p_{ij} are the following, for $p_0 = r$ and $p_1 = v$, with $r + v = 1$,

$$
p_{00} = rp, \; p_{01} = r(1-p), \; p_{10} = 0, \; p_{11} = (1-r)(1-p), p_{02} = 0,
$$
$$
p_{12} = (1-r)p.
$$

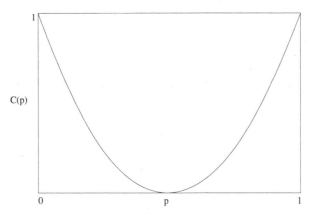

Figure 5.4 Graph for the capacity of the memoryless binary symmetric channel.

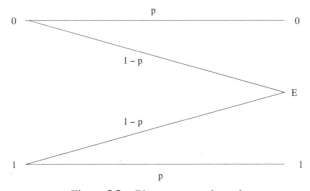

Figure 5.5 Binary erasure channel.

Substituting the probability values in eqn 5.64, one obtains

$$
\begin{aligned}
I(X;Y) = {}& [rp] \log \left(\frac{rp}{r^2 p} \right) \\
& + [(1-p)r] \log \left(\frac{(1-p)r}{(1-p)r} \right), \\
& + [(1-r)(1-p)] \log \left(\frac{(1-r)(1-p)}{(1-p)^2} \right) \\
& + [(1-r)p] \log \left(\frac{(1-r)p}{(1-r)^2 p} \right).
\end{aligned}
$$

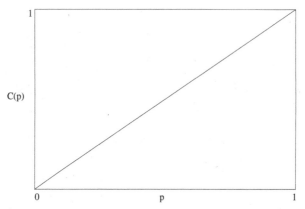

Figure 5.6   Graph of the capacity for the binary erasure channel.

Simplifying the terms in the expression gives

$$I(X;Y) = p\left[r \log r - (1 - r) \log (1 - r)\right] = p\, H(r), \quad (5.65)$$

in which $H(r)$ is the entropy function.

Because p is determined, $I(X;Y)$ is maximized by the choice of a value for r that produces the highest $H(r)$, that is, $r = 1/2$, for which $H(r) = 1$. Therefore, the capacity of the binary erasure channel is simply p. Figure 5.6 shows the graph for the capacity as a function of the probability p.

6

Computer Networks

"Cryptography is typically bypassed, not penetrated."

Adi Shamir

Computer networks have become ubiquitous and permeate the society. They are the main means of social communication, which includes voice, data, and video communication, and services, including electronic commerce, information search, distance learning, academic consultation, events, business, and malicious activities. A typical computer network, with the servers of network and users' machines, is illustrated in Figure 6.1.

Computer networks differ from telephone networks, which dominated the second half of the twentieth century, in the switching technique used

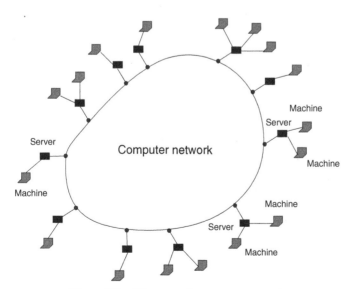

Figure 6.1 Elements of a computer network.

and in the transmission rates. They use packet switching, transmission rates, and variable topology. Information routing is performed at each node in the system.

The telephone networks, in turn, use exchanges with circuit switching, and they have a transmission rate limited to 64 kbits/s, for fixed telephony, and use a star topology, with the switches located in the center of the star (Hammond and O'Reilly, 1986).

The purpose of this chapter is to introduce the concepts of data flow, queue models, structure and topology, performance measures, traffic, channel capacity, latency, protocols, and more representative architectures of computer networks.

These concepts are quickly incorporated into the communication systems, in a way that its intrinsic data traffic characteristics need to be taken into account in the design of the systems. In particular, the information is essential to understand how common activities take place, such as sending files, searching for information, and telephony via the computer network, known as IP telephony.

6.1 Data Flow in Networks

The performance analysis of computer networks is related to the nature and characteristics of the data stream, which, in computer networks, is typically not uniform and random. The arrival times of messages, packages, or characters are random.

The time to process a message on the channel depends on the number of bits of the message, which can be random, so that the time also becomes a random variable. Efficiency, throughput, delay, and other parameters of interest are measures of how the network processes messages.

The queuing theory, which encompasses the theory of telephone traffic, provides an important model for the analysis of quantitative performance, correlating the most interesting computer networks (Hayes, 1986).

6.1.1 Capacity of a Channel Composition

The upper limit of the network's throughput is known as capacity. As discussed, the capacity of a symmetric binary channel (BSC), which adequately characterizes digital transmission on some real channels, is

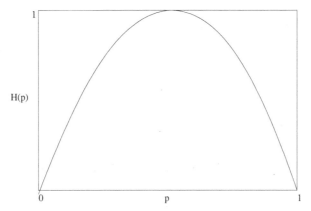

Figure 6.2 Graph of the binary entropy function.

given by

$$C = 1 - H(p) = 1 - p\log(p) - (1-p)\log(1-p),$$

in which $H(p)$ is the binary entropy function, illustrated in Figure 3.1, and repeated in Figure 6.2, in which p is the symbol exchange probability (crossover probability), which is related directly with the error probability of bit.

In the case of a computer network, such as the Internet, one can model, in a simplified way, the channel as a concatenation of channels in series, as shown in Figure 6.3. The composition of L channels in series is equivalent to a single BSC channel, with a probability of change of symbol exchange given by

$$p_L = \frac{1}{2}\left(1 - (1-2p)^L\right). \tag{6.1}$$

To obtain this probability, it is sufficient to note that the output of a composition of BSC channels is incorrect if, and only if, the transmitted symbol is changed an odd number of times, as it goes through the composition. An even number of inversions results in transmission without mistake.

The number of changes is calculated from the binomial distribution of the random variable X, for which

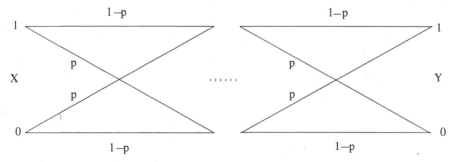

Figure 6.3   Model of a network formed by concatenating channels in series.

$$P(X \text{ odd}) = \sum_{i=1,3,5,\dots} \binom{L}{i} p^i (1-p)^{L-i}$$

$$= \frac{1}{2} \left(\sum_{i=1,3,5,\dots} \binom{L}{i} p^i (1-p)^{Li} + \sum_{i=0,2,4,\dots} \binom{L}{i} p^i (1-p)^{Li} \right)$$

$$+ \frac{1}{2} \left(\sum_{i=1,3,5,\dots} \binom{L}{i} p^i (1-p)^{Li} - \sum_{i=0,2,4,\dots} \binom{L}{i} p^i (1-p)^{Li} \right)$$

$$= \frac{1}{2} \left(\sum_{i=0}^{L} \binom{L}{i} p^i (1-p)^{Li} - \sum_{i=0}^{L} \binom{L}{i} (-p)^i (1-p)^{Li} \right)$$

$$= \frac{1}{2} \left(1 - (1-2p)^L \right).$$

Therefore, the capacity of the concatenation and L serial BSC channels is given by

$$C_L = 1 - H(p_L) = 1 - (-p_L \log(p_L) - (1-p_L)\log(1-p_L)). \quad (6.2)$$

If $p = 0$, then $p_L = 0$; so $H(P_L) = 0$ and the final capacity of the channel sequence is $C_L = 1$. If $p = 1$, then $p_L = 0$ or $P_L = 1$, depending on whether L is even or odd; so $H(P_L) = 0$ and a final capacity is $C_L = 1$.

The interesting data of the concatenation is that the final capacity of the channel concatenation tends to zero, when the number of channels in series grows without limit. If $0 < p < 1$, then $\lim_{L \to 0} p_L = 0.5$; so entropy tends to $H(P_L) = 1$ and the serial assembly capacity tends to $C_L = 0$.

In other words, a network formed by symmetrical binary channels in non-trivial series loses all transmitted information because its capacity converges to zero, as the number of channels tends to infinity.

Of course, this is true for a network that does not have transmission error correction. For M identical channels and independents in parallel, the total capacity, in the case where there is no interaction between the servers, is given by $C_M = M \cdot C$, that is, the addition of new channels increases the capacity in the same proportion. A network of communications is the composition of channels in series and in parallel, and the final capacity of that combination depends on the topology of each network.

This characteristic of information loss in a composition of channels in series, or waterfall, is more general and, in fact, is a theorem known as data processing inequality, and it was demonstrated by Philip Mayne Woodward, in 1955 (Abramson, 1963).

Suppose there is a probabilistic model described by the following Markov chain:

$$X \rightarrow Y \rightarrow Z,$$

that is, $p(X, Y, Z) = p(X)p(Y|X)p(Z|Y)$, in which X is independent of Z since Y is known. So it turns out that

$$I(X; Y) \geq I(X; Z). \tag{6.3}$$

To demonstrate ownership, you must define the conditional mutual information $I(X; Y|Z)$ between X and Y, given Z, as

$$\begin{aligned} I(X; Y|Z) &= H(X|Z) - H(X|Y, Z) \\ &= H(X|Z) + H(Y|Z) - H(X, Y|Z). \end{aligned} \tag{6.4}$$

Note that it derives from the unconditional property, which relates

$$\begin{aligned} I(X; Y) &= H(X) - H(X|Y) \\ &= H(X) + H(Y) - H(X, Y). \end{aligned} \tag{6.5}$$

By the chain rule for entropy, it is possible to decompose $I(X, (Y, Z))$ in two ways,

$$\begin{aligned} I(X; (Y, Z)) &= H(X) - H(X|Y, Z) \\ &= H(X) - H(X|Y) + H(X|Y) - H(X|Y, Z) \\ &= I(X, Y) + I(X, Z|Y) \end{aligned} \tag{6.6}$$

and, similarly,

$$I(X;(Y,Z)) = I(X;Z) + I(X;Y|Z). \tag{6.7}$$

Whereas $I(X;Z|Y) = 0$, depending on the independence between entry and exit, given the intermediate variable,

$$I(X;Z) + I(X;Y|Z) = I(X;Y).$$

Since mutual information is always non-negative, you get $I(X;Z) \leq I(X;Y)$. That is, the mutual information decreases, or does not change, with the composition of channels. Following the same reasoning, the capacity obtained with the combination of channels is always less than, or equal to, that of the constituent channels.

A computer network can have thousands of servers, interconnected by thousands of communication channels, usually with some kind of cooperation between the servers and users. This makes capacity analysis a complex task. Therefore, designers use simpler parameters to be specified, such as flow (throughput) and latency.

6.1.2 Stationary Data Flow

Several properties related to stationary flow packets or messages can be derived regardless of the distributions of the parameters that define the flow. In the analysis of the data flow, it is assumed that the network conserves the messages, in the sense that they cannot be created, destroyed, or modified by the network.

Messages can only flow into or out of outside the perimeter of the network, or can remain stored for a certain time on the network. The time spent in the network corresponds to the sum of the processing times of the packets by the servers. On each server, the message needs to be opened, eventually corrected, and then routed to the next destination.

If the average rate of entry into the network exceeds the rate of exit, the number of stored messages is constantly increasing. On the other hand, if the average rate is higher than the input rate, the number of stored messages decreases to zero.

From these considerations, one concluded that, for a stable operation of network in the steady state, the input and output rates must be equivalent.

Consider $\alpha(t)$ as the number of incoming packages and $\delta(t)$ the number of network outgoing packets in a certain time interval $(0,t)$. The difference between these quantities, $N(t)$, represents the increase in the number of messages stored on the network in the given interval

$$N(t) = \alpha(t) - \delta(t). \tag{6.8}$$

The input rate, in a time interval t, is defined as

$$\lambda_t = \frac{\alpha(t)}{t}. \tag{6.9}$$

Another measure of interest is the total time that all messages spend on the network

$$\gamma(t) = \int_0^t N(x)dx = \int_0^t \alpha(x) - \delta(x)dx. \tag{6.10}$$

Proceeding in this way, it is possible to find the average number of messages, N_t, on the network in the range $(0, t)$

$$N_t = \frac{1}{t}\int_0^t N(x)dx = \frac{\gamma(t)}{t}. \tag{6.11}$$

The average time spent by any message that has entered in the range $(0, t)$ is given by

$$T_t = \frac{\gamma(t)}{\alpha(t)}. \tag{6.12}$$

The previous formulas can be combined, taking the limit of each term, assuming that they remain finite, providing

$$N = \lambda T, \tag{6.13}$$

which is known as Little's law, in honor of Professor John Little, and establishes that the number of messages stored in a network is equal to the product of the rate of incoming messages to the network by the average time that these messages remain on the network, for a stationary system (Little, 1961).

Considering the network of Figure 6.4, equipped with a unit of storage or buffered and an output channel with C bits per second, it is assumed that the λ messages 'e entry fee per second, that the average length of messages é $1/\mu$ bits per message and the traffic intensity, or time the channel is busy, is given by ρ (Alencar, 2012a).

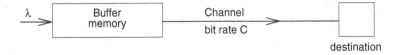

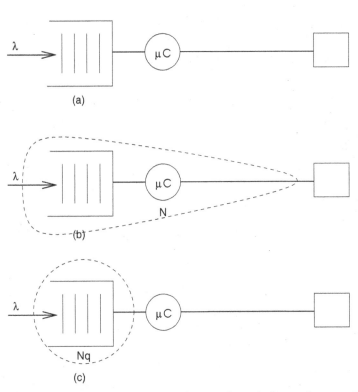

Figure 6.4 The network's theoretical model. The top figure indicates the network as the composition of a memory and a channel of communications with C capacity. The lower figures represent, respectively, (a) the model in the network note, (b) the model with a region including the buffered and the channel, and (c) the model with the region that includes only the memory. Adapted from (Alencar, 2012a).

The average transmission or processing time is $1/\mu C$ seconds. This implies that μC messages processed per second, so that the output rate can be expressed by

$$\lambda = \rho\mu C \tag{6.14}$$

or

$$\rho = \frac{\lambda}{\mu C} \qquad (6.15)$$

This indicates that the traffic intensity equals the arrival rate, divided by the channel transmission rate. In order for there having no messages left in the network, which would lead to unstable behavior, the arrival rate should be less than or equal to the channel average processing rate, and then $0 \le \rho \le 1$.

Little's law can be applied to region which includes the buffer and the channel, providing

$$N = \lambda T. \qquad (6.16)$$

If the region involves only the buffer, which stores N_q messages in average for a period of W seconds, one has

$$N_q = \lambda W. \qquad (6.17)$$

As T is the sum of the average delays in the buffer and in the channel, one obtains

$$T = W + \frac{1}{\mu C} \qquad (6.18)$$

A relationship can be obtained between N and N_q from the previous equation, multiplying the equality by λ, which results in

$$N = N_q + \rho. \qquad (6.19)$$

This equation shows that the number of messages stored in the network is equal to the average quantity stored in the buffer, N_q, plus rho, which must represent the number of messages stored in the channel.

6.2 Queue Models

Computer networks generally start out structured but typically evolve in an amorphous way, as computational demands appear. The network manager, in addition to dealing with keeping the system running and its security, needs to take care of network optimization, which involves knowledge of queuing theory (whittle.2007; garcia.2000)

Queuing theory provides adequate models for the quantitative analysis of system and network performance, or even user behavior, and allows you to correlate important parameters. The usual notation indicates the premises for the input process, the server process, and the number of output channels (Kleinrock, 1975).

The characteristics that describe a queue, for simplicity, are described using the notation of the English mathematician and statistician David George Kendall (1918–2007), proposed in 1953 (Kendall, 1953), composed of a series of symbols of the form A/B/C/D/E/Q, in which:

A – Distribution of times between successive arrivals.

B – Distribution of service times, or attendance.

C – Number of servers, or service stations.

D – Physical capacity of the system.

E – Size of the population.

F – Service discipline.

The most used probability distributions in queuing theory are:
- Exponential (M).
- Uniform (U).
- Arbitrary or general (G).
- Erlang type k (E_k).
- Hyperexponential (H_k).

Some of the most common service disciplines in queuing theory are:

FIFO – First In, First Out, that is, the first to enter is the first to leave.

FCFS – First Come, First Served, that is, the first to arrive is the first to be served.

LCFS – Last Come, First Served, that is, the last to arrive is the first to be served.

LIFO – Last In, First Out, that is, the last to arrive is the first to leave.

The following examples illustrate the use of Kendall's notation:

$M/G/1$ – In the case of an entry process being Poisson, without restriction on the server process and only one output channel.

$M/M/20$ – If the entry process is from Poisson, as well as the server process, with 20 output channels.

$M/D/1$ – If the entry process is from Poisson, but the server is deterministic, with only one output channel.

$M/E_k/30/200$ – If the entry process is from Poisson, the server is Erlang type k, with 30 channels of output and capacity for 200 clients.

$M/M/2/\infty/\infty/\text{FIFO}$ – If the entry process is from Poisson, as well as the server process, with two output channels, unlimited capacity, infinite population, and service discipline in which the first to enter is the first to leave.

6.2.1 Markov Model

The Markov model shown in Figure 6.5 serves as a basis for the analysis traffic on computer networks. A discrete-time Markov chain is used to model a process stochastic set in honor of mathematician Andrei Andreyevich Markov. For this chain, the states prior to the current one are not relevant to the prediction of future states, as long as the current state is known.

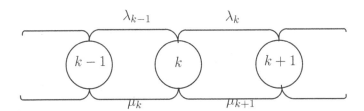

Figure 6.5   Markov model of birth and death.

Known as the model of birth and death, transitions occur only between states adjacent. For example, from the state k, you can go only to $k + 1$ or $k - 1$ with some probability. This reflects the fact that the likelihood of more than one user entering the system at the same time is negligible. Using the model, it is possible to calculate the steady state probabilities (Kleinrock, 1975).

The transition matrix probabilities $\mathbf{P} = \{p_{ij}\} = \{p(y_j|x_i)\}$ defines the dynamics of the model. The transition probabilities are obtained from the Markov model, in which λ_k and μ_k are the birth and death parameters.

$$\mathbf{P} = \begin{bmatrix} p(y_1|x_1) & p(y_2|x_1) & \cdots & p(y_N|x_1) \\ p(y_1|x_2) & p(y_2|x_2) & \cdots & p(y_N|x_2) \\ \cdots\cdots\cdots\cdots\cdots\cdots \\ p(y_1|x_M) & p(y_2|x_M) & \cdots & p(y_N|x_M) \end{bmatrix} \tag{6.20}$$

The Markov chain reaches a steady state after a certain number of iterations. The probabilities of the steady state, $\Pi = \{\, pi_k | k = 1, 2, 3...\}$, can be calculated using one of the known techniques (Kleinrock, 1975; Adke and Manjunath, 1984). Each k state defines the number of users, packages, or other objects in the system.

Two cases of application of the Markov chain are presented. Initially, the problem arises when the birth and death parameters are constant for any state. It can be determined that $\lambda_k = \lambda$ and $\mu_k = \mu$.

The first case, which illustrates the operation of a traditional computer network, connected by wires or cables, which operates with defined flow parameters, produces a geometric distribution for the probabilities

$$p_k = (1 - \rho)\rho^k \ \ k = 0,1,2, ..., \text{ for } \rho < 1, \tag{6.21}$$

in which $\rho = \lambda/\mu$ is usually called the use of the system. Figure 6.6 illustrates the geometric probability distribution as a function of the state of the k system.

For the geometric distribution, the statistical average is given by $\rho/(1-\rho)$ and the variance by $\rho/(1 - \rho)^2$. The probability of finding more than L users at a given time in the system is ρ^{L+1}.

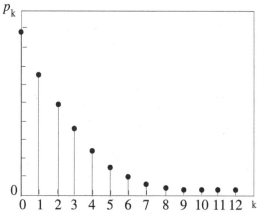

Figure 6.6   Geometric probability distribution, depending on the state of the k system.

Then, the problem that the system server is efficient is calculated, and it reacts to the increased data flow in the system. This is equivalent to the case when users drop out in function, for example, the delay in completing a connection.

This Markov model exemplifies the operation of a wireless computer network, for example, as in the case of the interconnection of notebooks, tablets, or cell phones, in which there is a competition for access to the channel. For this case, we obtain $\lambda_k = \lambda/(k+1)$ and $\mu_k = \mu$, in which λ and μ are fixed probabilities.

The solution for the second case generates a distribution from Poisson

$$p_k = \frac{\rho^k}{k!} e^{-\rho} \ k = 0,1,2,\ldots \tag{6.22}$$

with month and variance given by ρ. Figure 6.7 illustrates the geometric probability distribution as a function of the state of the k system.

6.2.2 Solution for a Stable Network

The formula that relates the probability of k existence packets at a given time on a network is given by the solution of the following equations, which depends on the configuration assumed for the system (Kleinrock, 1975)

$$p_k = p_o \prod_{i=0}^{k-1} \frac{\lambda_i}{\mu_{i+1}},$$

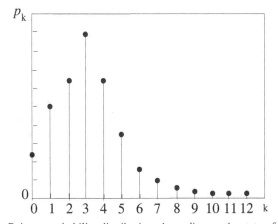

Figure 6.7 Poisson probability distribution, depending on the state of the k system.

$$p_o = \cfrac{1}{1 + \sum\limits_{k=1}^{\infty} \prod\limits_{i=0}^{k-1} \frac{\lambda_i}{\mu_{i+1}}},$$

$$N = \sum_{k=0}^{\infty} k p_k,$$

$$\sigma_N^2 = \sum_{k=0}^{\infty} (k - N)^2 p_k,$$

$$N = \lambda T,$$

in which:

- p_o – probability that there will be no packages in the system;
- p_k – probability of having k packages in the system;
- λ_i – package arrival fee to state i;
- μ_{i+1} – departure fee for state packages $i + 1$;
- N – average number of packages in the system;
- σ_N^2 – variance of the number of packages in the system;
- T – average waiting time in line.

These equations serve as a starting point for calculating several queuing theory problems and for other solutions shown below in this chapter.

6.2.3 Constant Arrival Rate System

In this type of system, the arrival and departure rates of calls are and can be described, in terms of the coefficients, as follows:

$$\lambda_k = \lambda, k = 1,2,3,4, ...$$
$$\mu_k = \mu, k = 1,2,3,4, ...$$

The probabilities of k packages and no package are given by:

$$p_k = p_o \prod_{i=0}^{k-1} \frac{\lambda}{\mu},$$

$$p_k = p_o \left(\frac{\lambda}{\mu}\right)^k,$$

$$p_o = \cfrac{1}{1 + \sum\limits_{k=1}^{\infty} \prod\limits_{i=0}^{k-1} \frac{\lambda_i}{\mu_{i+1}}},$$

$$p_o = 1 - \frac{\lambda}{\mu},$$

with the restriction that $\frac{\lambda}{\mu}$ must be less than 1. Making $\rho = \frac{\lambda}{\mu}$, one obtains

$$p_k = (1 - \rho)\rho^k$$

and

$$p_o = (1 - \rho).$$

The number of packages in the system is the sum of the product of the number of packets in the network and is determined by the probability of the occurrence of this number of packages. Mathematically, this can be represented by

$$N = \sum_{k=0}^{\infty} k.p_k.$$

After some mathematical manipulations, remembering that

$$\frac{d}{d\rho} \sum_{k=1}^{\infty} \rho^k = \sum_{k=1}^{\infty} k\rho^{k-1}$$

and that

$$\sum_{k=1}^{\infty} \rho^k = \frac{1}{1 - \rho},$$

one has

$$N = \frac{\rho}{(1 - \rho)}.$$

Figure 6.8 illustrates the growth in the number of packets on a network, with the coefficient ρ.

The variance in the number of packages in the system é is given by

$$\sigma_N^2 = \sum_{k=0}^{\infty} (k - N)^2 p_k,$$

$$\sigma_N^2 = \frac{\rho}{(1 - \rho)^2}.$$

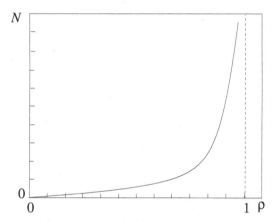

Figure 6.8   Growth in the number of N packets on the network with ρ.

The results described do not depend on the parameters λ and μ individually, but only on the quotient between them, the ratio between arrival, and processing fees.

The average waiting time in the queue is given by the quotient between the average number of packages and the average rate of arrival of packages to the system. Therefore,

$$T = \frac{N}{\lambda} = \frac{1/\mu}{(1 - \rho)}.$$

Figure 6.9 illustrates the increase in latency in a network, as the ρ coefficient increases.

When $\rho = 0$, the T latency corresponds to the expected service time for the first package, being equal to $1/\mu$. This average time depends on the parameter μ individually, contrary to the previous ones.

Note that, as ρ tends to the unit, that is, the rate of packets that arrives on the network approaches the processing rate of the network's servers, the average number of packets in the system and the average waiting time in the queue tend to grow in an unlimited way. This type of behavior when ρ tends to 1 is characteristic of almost all queuing systems encountered.

In general, this feature is exploited by hackers for an attack known as denial of service (DoS), where the network receives a flood of packets so that

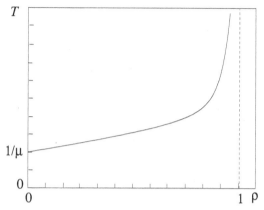

Figure 6.9 Increased network latency due to ρ.

the arrival rate approaches processing rate, and the number of packets on the network tends to infinity.

Another interesting amount to be analyzed is the probability that there must be at least k packages in the system. This probability is given by

$$P[x \geq k \text{ in the system}] = \sum_{i=k}^{\infty} p_i$$

$$= \sum_{i=k}^{\infty} (1 - \rho)\rho^i = \rho^k,$$

for ρ less than 1, since $\sum_{i=k}^{\infty} \rho^i$ only converges if $\rho < 1$.

6.2.4 Efficient Server Model

This type of system can be interpreted in two ways: as a server responsible for accelerating your service fee as the number of packets in the queue increases or as a new server available for each incoming packages. The model can be described by the following parameters:

$$\lambda_k = \lambda \quad k = 1,2,3,4,\dots$$
$$\mu_k = k\mu \quad k = 1,2,3,4,\dots$$

The functions found are given by

$$p_k = \frac{(\lambda/\mu)^k}{k!} e^{-\frac{\lambda}{\mu}},$$

$$p_o = e^{-\frac{\lambda}{\mu}},$$

$$N = \frac{\lambda}{\mu},$$

$$T = \frac{1}{\mu}$$

from Little's equation.

It can be noted that the system with an efficient server is equivalent to a system with discouraged arrivals when $\alpha = \lambda$.

6.2.5 Model with Discouraged Arrivals

In general, it is common to consider only the fixed network of computers interconnected by cables of wires, or optical, in the solution of problems related to the latency and number of packets stored. However, the access network, typically wireless, is also important and includes the equipment used for the connection, such as cell phones, notebooks, and tablets.

In this type of network, competition is established between users to gain access to available channels. This competition changes the scenario used for mathematical modeling. The fundamental change is to make the arrival rate dependent on the number of users (or packages) on the network.

It is now considered an access system in which the entry of users tends to be discouraged, as the number of packages increases in system. To model this system, the coefficients must be assigned as follows:

$$\lambda_k = \frac{\alpha}{k+1} \quad k = 1,2,3,4, \ldots$$
$$\mu_k = \mu \quad k = 1,2,3,4, \ldots$$

It is assumed that there is a lack of incentive for new arrivals, as the number of packages in the system increases. For this system, one obtains

$$p_k = p_o \prod_{i-0}^{k-1} \frac{\frac{\alpha}{i+1}}{\mu},$$

$$p_k = p_o \left(\frac{\alpha}{\mu}\right)^k \frac{1}{k!},$$

$$p_o = \frac{1}{1 + \sum\limits_{k=1}^{\infty} \prod\limits_{i-0}^{k-1} \frac{\lambda_i}{\mu_{i+1}}},$$

$$p_o = e^{-\frac{\alpha}{\mu}},$$

$$\rho = 1 - e^{-\frac{\alpha}{\mu}},$$

$$N = \sum_{k=0}^{\infty} k \cdot p_k,$$

$$N = \frac{\alpha}{\mu},$$

$$T = \frac{N}{\lambda},$$

$$T = \frac{\alpha}{\mu^2 \left(1 - e^{-\frac{\alpha}{\mu}}\right)},$$

$$P[x \geq k \text{ in the system}] = \sum_{i=k}^{\infty} p_i$$

to be calculated.

This solution is equivalent to the one obtained for the efficient server, that is, different assumptions can lead to similar mathematical models.

6.2.6 Models of Queues M/G/1, M/M/1, and M/D/1

The $M/G/1$ model is used as a starting point for analysis. With respect to the service process, it is convenient to consider the message processing as a sequence of variables independent and identically distributed.

For the Poisson entry process, the number of arrivals at any time is statistically independent of the number of messages arriving at any other interval are not superimposed.

The probability of k arriving in a time interval T is given by

$$P\{k\} = \frac{(\lambda T)^k e^{-\lambda T}}{k!}, \quad k = 0, 1, 2, \ldots. \tag{6.23}$$

The message arrival rate, that is, the average number of message arrivals at $T = 1$ second, is *lambda* messages per second. Analyzing the $M/G/1$ model, it is concluded that the number medium of messages on the network is given by

$$N = \frac{2\rho - \rho^2 + \lambda^2 \sigma_Y^2}{2(1 - \rho)} = \rho + \frac{\rho^2 + \lambda^2 \sigma_Y^2}{2(1 - \rho)}, \tag{6.24}$$

in which σ_Y^2 is the variance of the processing time of the message, Y.

The application of Little's law allows to obtain an expression for the average time of the message in the system (Alencar, 2012a),

$$T = \frac{1}{\mu C} + \frac{\rho + \lambda \mu C \sigma_Y^2}{2\mu C(1 - \rho)}. \tag{6.25}$$

These are the formulas of the average value of Pollaczek–Khinchine, which show the growth in the number of packets in the N network, for typical values of network fees and capacity. The model $M/M/1$ is a special case of the previous model. In this case, the processing time of the message is described by a Poisson process.

Therefore, it can be demonstrated that the variance of the time of processing is given by $\sigma_Y^2 = (1/\mu C)^2$ and, thus, the number of messages in the system becomes

$$N = \frac{\rho}{1 - \rho}. \tag{6.26}$$

As a result, the average delay time on the route between the buffer and the channel is given by

$$T = \frac{\rho}{\lambda(1 - \rho)} = \frac{1}{\mu C(1 - \rho)}. \tag{6.27}$$

The formulas for N_q, the average number of messages stored in the buffer, and W, the average time in the buffer, can be obtained, respectively, by

$$N_q = \frac{\rho^2}{1 - \rho} \tag{6.28}$$

and

$$W = \frac{\rho}{\mu C(1 - \rho)}. \tag{6.29}$$

It should be noted that the model $M/D/1$ is also a special case of the model $M/G/1$, in which the message processing time is set as invariant and denoted by $1/\mu C$.

As the variance in this case is zero, one can write

$$N = \rho + \frac{\rho^2}{2(1-\rho)} = \frac{\rho(2-\rho)}{2(1-\rho)} \qquad (6.30)$$

and

$$T = \frac{1}{\mu C} + \frac{\rho}{2\mu C(1-\rho)} = \frac{2-\rho}{2\mu C(1-\rho)}. \qquad (6.31)$$

As a consequence,

$$N_q = \frac{\rho^2}{2(1-\rho)} \qquad (6.32)$$

and

$$W = \frac{\rho}{2\mu C(1-\rho)}. \qquad (6.33)$$

The curves in Figure 6.10 show the number of packets in the network N versus ρ, for an $M/D/1$ queue, compared with similar results for an $M/M/1$ queue (Alencar, 2012a).

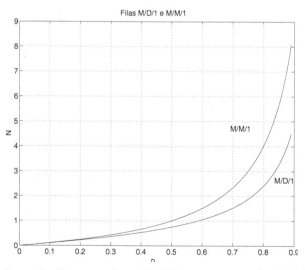

Figure 6.10 Curves for N versus ρ, for an $M/D/1$ queue, compared to the $M/M/1$ queue.

6.3 Local Area Networks

Computer networks often make use of efficient communication channels and take advantage of your potential for resource sharing among users (Moura et al., 1986; Giozza et al., 1986). The communication of processes takes place through the network between its edge nodes.

A generic computer network can be defined as a set of autonomous computers interconnected so that no machine completely controls the others. Computer networks have been developed to enable the sharing of facilities and the distribution of resources among users. The function of the network is to provide a means by which the computers connected to the network can exchange data.

6.3.1 Definitions and Functions

From the user's point of view, computer networks have become mostly for the exchange of information, for the purchase of products and services, and for carrying out works of various kinds. There are several types of computer networks. A geographic classification is generally used (Tanenbaum, 1989):

- local area networks;
- wide area networks.

Wide area networks, such as the Internet, cover large geographic areas, often extending nationally or globally. Local area networks, such as Fibernet, Ethernet, or Novell, are characterized by restricted physical dimensions, usually covering a few hundred meters.

The difference in physical dimensions has an effect pronounced on the structure of networks. The local network, due to its limited physical extension, may be owned by network users. In this way, it is possible to use dedicated high-speed and reliable channels.

The typical means of transmission on local computer networks are twisted pair, optical fiber link, 50-Ω coaxial cable (for basic band transmission), and 75-Ω coaxial cable (for broadband transmission). Wireless networks are also used, such as Wi-Fi (IEEE 802.11), whose IEEE 802.11n standard employs transfer rates from 65 to 600 Mbit/s, with transmission based on MIMO-OFDM, operating in the frequency range of 2.4 or 5 GHz (Gast, 2005).

Long distance networks, for economic reasons normative, used telephone links, which resulted in links with lower speeds. With the launch of long-term optical networks distance network, which includes OPGW networks,

transmission over high voltage lines, from electricity, transmission rates have risen considerably.

Wireless networks are also formed with WiMAX technology (IEEE 802.16), which stands for worldwide interoperability for microwave access, and offers fixed, nomadic, portable or mobile connectivity, without the need for direct sight with a base station. For a typical scenario, with distances from 3 to 10 km, WiMAX presents a transmission rate of 40 Mbit/s per channel (IEEE, 2004).

Four physical layer standards have been specified for the metropolitan WiMAX network: single carrier (SC), single carrier a (SCa), orthogonal frequency- division multiplexing (OFDM), and orthogonal frequency-division multiple access (OFDMA). However, cellular mobile telephony with long-term technology evolution (LTE) eclipsed WiMAX.

6.3.2 Applications for Local Area Networks

One of the most versatile applications for local networks is the office automation, in which the local network interconnects work stations, computing equipment, and printers (Soares et al., 1995).

In an academic or business environment, the local network can provide the interconnection of work stations for development of computer-aided programs or projects.

Point-of-sale systems can have their equipment, such as cash registers and optical readers, in addition to consultation of files interconnected in network.

Local area networks can also be used in systems banks, libraries, laboratories, computational facilities in universities, and several other applications that require communication between computers. Figure   6.11 illustrates a generic point-to-point (P2P) local area network.

Local networks implement data transfer in discrete units called packets. A package can vary from a few bits to thousands of bits according to each particular protocol. Eventually, there is a general server on the network, as in the case of the model client-server of Figure 6.12.

6.3.3 Network Structure and Topology

The nodes of a network can be connected in several ways. The unstructured or amorphous topology is common in long-haul networks distances, which grow due to the disorderly demand for services but uncommon on local networks.

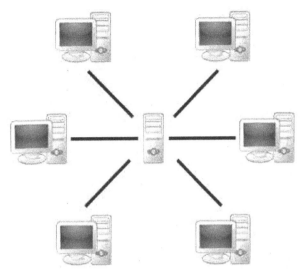

Figure 6.11   Generic local peer-to-peer network (Wikimedia Commons license in public domain).

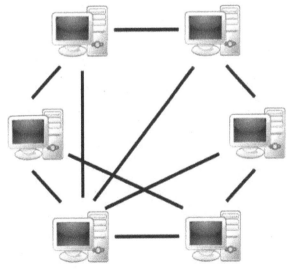

Figure 6.12   Generic local client-server network (Wikimedia Commons license in public domain).

The storage and retransmission operation is typical of unstructured networks, which allows flexibilization in the positioning of the nodes and in the communication between them. Local area networks have a combination of requirements that make the storage and retransmission operation unnecessary.

In these networks, efficient communication between devices is the goal. Therefore, the delay and cost associated with routing messages between nodes are considered excessive.

Unstructured networks generally need to perform the error control at each link due to the insertion of noise in the path. Error control between source and recipient is enough for local networks. Local networks frequently use the broadcasting of messages, which is unusual in long-distance networks.

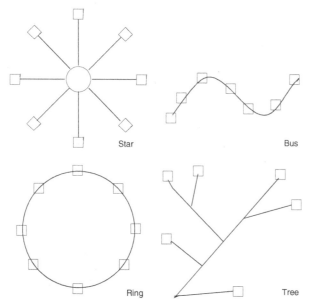

Figure 6.13 Structures for typical networks. Adapted from (Alencar, 2012a).

Some common topologies for local networks have characteristics that are appropriate for the provided services, such as the star, ring, and bus. The tree structure is a generalization of the bus.

For bus topology, this station sends the message in both directions. Each station, or note, has the ability to read the package heading, and there are buffers to store messages for which the node is the destiny.

In the ring network, communication is typically in one direction. Routing is, therefore, unnecessary. The ring topology uses the central node as a traffic redirector, as s in a telephone exchange. The links must be of the full-duplex type.

The tree network structure is common for pay-TV (CATV), which uses coaxial cable, optical cable, or a hybrid optical-coaxial system, and access by frequency division. Local networks can use the CATV cabling, or use the electrical network, with the technology power line communications (PLC) (Alencar et al., 2006).

6.3.4 Channel Access Techniques

There is a number of different access techniques that are used for different topologies. The choice of the technique of access is a determining factor in network performance. Techniques can be classified into four major categories: fixed allocation, random allocation, allocation on demand, and adaptive allocation.

In the fixed allocation, resources are placed at disposition of each station in a predetermined way. Two common schemes in this category are multiple access by multiple division into frequency (FDMA) and division into time (TDMA). At the opposite end to the fixed allocation strategy, there are the random allocation methods. The simplest method is known as pure ALOHA, in which each station transmits as soon as it is ready.

The adaptive or on-demand allocation methods require a control mechanism, operating in real time, to allocate the channel capacity to all stations in an optimum or quasi-optimum manner.

The on-demand allocation methods are still classified as central or distributed control. Algorithms to monitor systems illustrate central control, while ring networks use distributed control.

The adaptive allocation methods try to refine the random of on-demand allocation protocols. The adaptive protocols usually perform contention resolution, provided that collisions are allowed, under low loading, to minimize access times. For high loading, however, the procedures adjust for some form of polling or TDMA.

A characteristic of these methods is that some estimate of the load on the network is required by all stations.

6.3.5 Traffic on the Network

In a computer network, the nature of the traffic offered by the user devices is an important factor in evaluating performance (Alencar, 2012a).

However, the traffic offered to the network is variable and random, because it depends on many factors. Peak data rates range from 100 bits/s for security systems, telemetry, and network wireless sensors, at 100 Mbits/s for video transmission.

The definition of distribution for the stationary state is a difficult problem because of the significant differences between averages taken at distinct time intervals.

The data in Figure 6.14, for example, suggest an exponential distribution for the arrival times, and this distribution is often used as a basic model for research and development of networks

$$p_T(t) = \beta e^{-\beta t} u(t), \tag{6.34}$$

in which $u(\cdot)$ represents the unit step function and α is a parameter to be estimated.

The distribution of packet lengths over an Ethernet network experiment has a bimodal character. Smaller packages occur depending on the interactive traffic (bursty), while larger packages represent the result of the transfer of large volumes of data.

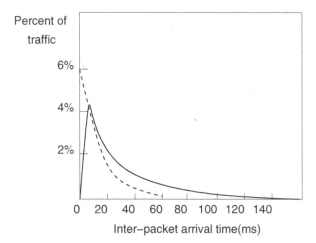

Figure 6.14 Distribution of arrival time between packets on an Ethernet network. Adapted from (Hammond and O'Reilly, 1986).

6.3.6 Performance Measures

An important performance measure, from the point of view of the user, is the network response time, that is, the time to correctly transmit a packet from one user to another, with the confirmation. Note that both network delays and user/station link delays can dominate the response time between users. If the network delays are significant, the choice of the best access technique is very important.

At the network level, a specific measure is the throughput, which measures the number of bits/s or packets/s, which can, on average, be processed by the network. If the average input rate for one part of the network is λ packets/s, the channel transmission rate is R bits/s and there is $\bar{X}$ bits/package on average; therefore, the normalized throughput is given by

$$S = \frac{\lambda \bar{X}}{R}. \tag{6.35}$$

Throughput is used in the design and performance evaluation of computer networks because the capacity, based on information theory, is complicated to calculate for a complex network. But, it must be kept in mind that the upper limit for the transmission rate in a communications system is its capacity, calculated as the maximum of the information mutual support for all probability distributions of the input alphabet.

The average transfer delay, T, is defined as the time between the arrival of the last bit of a packet to a network station and the delivery of the last bit of that packet to the destination. The normalized average delay, in relation to the average time of transmission in the channel, is given by

$$\hat{T} = \frac{T}{\bar{X}/R} = \frac{RT}{\bar{X}}. \tag{6.36}$$

The delay, or latency, of the network is an important parameter when dealing with voice transmission in real time, as in the case of voice over IP (VoIP), which is the technology used in applications like WhatsApp and Skype.

The computation of the total delay, T, or network latency, which is a random variable, is done using properties of probability theory. Consider that each partial delay time, typically represented by the processing time of the message by each individual server, plus the time traffic on the corresponding link, has exponential probability distribution.

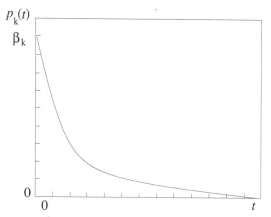

Figure 6.15 Probabilistic model for the delay time in a computer network.

$$T = \sum_{k=1}^{M} t_k, \tag{6.37}$$

in which M is the number of servers in the message path, and each particular delay time has distribution

$$p_k(t) = \beta_k e^{-\beta_k t} u(t), \tag{6.38}$$

in which $u(\cdot)$ is the unit step function and β_k represents the parameter it regulates the probability distribution for the time t_k. The distribution is illustrated in Figure 6.15.

The sum of independent random variables leads to the convolution of two respective distributions of probability. Therefore, the probability distribution of the total time is given by

$$p_T(t) = p_1(t) * p_2(t) * \cdots * p_M(t), \tag{6.39}$$

in which the convolution operation between two functions $f(t)$ and $g(t)$ is defined as

$$f(t) * g(t) = \int_{-\infty}^{\infty} f(\tau)g(t - \tau)d\tau. \tag{6.40}$$

Example: For two servers, or two subnets, $M = 2$, the calculation can be done by directly using Formula (6.40),

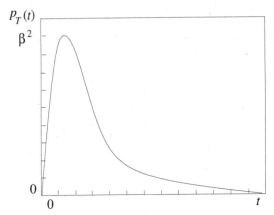

Figure 6.16   Probabilistic model for combined latency in two subnets with same delay time.

$$p_T(t) = \int_{-\infty}^{\infty} p_1(\tau)p_2(t-\tau)d\tau$$

$$= \int_{-\infty}^{\infty} \beta_1 e^{-\beta_1 \tau} u(\tau)\beta_2 e^{-\beta_2(t-\tau)}u(t-\tau)d\tau$$

$$= \beta_1\beta_2 e^{-\beta_1 t} \int_0^t e^{-(\beta_2-\beta_1)\tau}d\tau$$

$$= \frac{\beta_1\beta_2}{\beta_1-\beta_2}\left(e^{-\beta_2 t} - e^{-\beta_1 t}\right)u(t).$$

If the processing times are the same for both servers, that is, the parameters $\beta_1 = \beta_2 = \beta$, the solution converges to

$$p_T(t) = \beta^2 t e^{-\beta t}u(t).$$

The solution for the combined latency of two servers, or two subnets, is represented in Figure 6.16, which resembles the distribution of the arrival time between packets on an Ethernet network, illustrated in Figure 6.14.

The general case, for M servers with different processing times, produces the following probability distribution for the total time (Akkouchi, 2008).

$$p_T(t) = \sum_{k=1}^{M} \frac{\beta_1 \cdots \beta_M}{\prod_{i=1,i\neq k}^{M}(\beta_i - \beta_k)} e^{-\beta_k t}u(t). \tag{6.41}$$

The probability distribution allows to calculate the average total network delay, latency, and also the deviation around this delay, which indicates the jitter. Latency is a measure of time that a package takes to get from the source to the destination. The jitter is the stochastic variation of delay in data delivery, that is, the temporal variation between successive data packets. The jitter limits the minimum separation between bits, which affects the baud rate on the network.

7

Network Protocols and Architecture

"A little bit of math can accomplish what all the guns and barbed wire can't: a little bit of math can keep a secret."

Edward Snowden, Permanent Record

The word protocol comes from the Greek word *protókollon*, meaning leaf placed in front, and came to English in medieval Latin *protocollu*, with the usual meaning of the word protocol, that is, a set of rules or criteria fulfilled in a given activity, whether in the execution, evaluation, or acceptance of services and materials.

In informatics and engineering, protocol means the set of rules that makes possible the execution of a program efficiently and without errors, or the exchange of information between servers of a computer network, for sending information. Nowadays, protocols in evidence are those associated with the Internet, the worldwide network that connects most of Earth's population. But it was not always so (Alencar, 2010).

7.1 What the World Was Like Without the Internet

Incredible as it may seem, there was a time when there was Internet. At that time, when you wanted to set up a meeting, it was necessary to call. To announce a birthday or birth, it was common use to send a telegram. People would even leave their rooms, in companies, to talk to colleagues about a work topic, or just to make small talk (Alencar, 2008a).

Of course, some companies started out in the business of exchanging posts. In the 1970s, there was the Telex service, provided by the European Post, Telephone and Telegraph companies (PTT), the American Telephone and Telegraph Corporation (AT&T), the Brazilian Telecommunications

123

Company (Embratel), and other public companies. They formed an international system of written communications, which prevailed until the 1990s, and consisted of a worldwide network with a numeric addressing plan, with terminals that could send written messages to other terminals.

The Telex had some limitations, as it did not have all the characters used in the most languages, such as accents, and all the letters were capitalized. The older machines had no random access memory (RAM). Therefore, the messages were pre-recorded by mechanical means, typically with perforated paper tape, and then transmitted. Incoming messages were printed on continuous rolls of paper (Alencar, 2011b).

The service worked in some countries, until the past decade, despite the number of falling subscribers due to the introduction of electronic mail. The terminals resembled electric typewriters and there was a guarantee of immediate delivery, with terminal authentication, which was important for the companies that hired the service (Alencar, 2011c).

In Brazil, Embratel already had an information packet exchange service in the early 1980s. It was known as Projeto Ciranda and served only its employees. Then Cirandão was launched, to the public access, but few people took notice. The marketers of company changed its name to STM-400, but it did not have much effect on the market. The company missed the target audience. The company should have bet on the university market.

Regarding the use of data communication networks, a significant event was the decision of the National Science Foundation (NSF) of the United States, in 1985, to invest in setting up networks to serve the academic and research community. In 1986, articles were published describing the networks used by the academic and research community in the USA, with emphasis on the Bitnet network, which has been in operation since 1981, and for the National Science Network Foundation (NSFNET), created in 1986.

Bitnet, also known as Because It's Time Network, was a large computer network, which carried electronic mail messages. It used technology developed by IBM, mainly the protocol Network Job Entry (NJE), and gave rise to the listserv program for maintenance of lists of debates. Its appeal was its simplicity of adhesion and operation, mainly for institutions that had an IBM computer (Alencar, 2008b).

Bitnet was managed by the Corporation for Research and Educational Networking (CREN), from Washington, USA, and was used to provide electronic mail and file transfer between large computers of educational and research institutions in North America, South America, Europe, and Japan. It reached more than 2-500 universities and institutes worldwide.

NSFNET, which would be part of the Internet, used the family of protocols TCP/IP, developed from Defense Advanced Research Projects Agency (DARPA). This protocol would allow several applications via the network, especially the interactive use of remote computers (Telnet), the transfer of files, known as File Transfer Protocol (ftp), and, from the decade 1990s, interactive consultation of information bases on the World Wide Web (www), in addition to electronic mail.

Due to the importance for the academic community of the use of computer networks, in the 1980s, several projects in this direction, at the National Scientific Computing Laboratory (LNCC) of CNPq, at the São Paulo State Research Support Foundation (Fapesp), and at the Federal University of Rio de Janeiro (UFRJ).

The first meeting to discuss the establishment of a national network for researchers, with access to international networks, was held at the University of São Paulo (USP), in October 1987. The meeting was attended by research institutions, development agencies, and Embratel. The network aimed at allowing broad access for members of the research community to Bitnet, using the facilities of the LNCC, through dial-up or through the National Network of Embratel's Packages (Renpac).

The message exchange service for university students in Brazil started in mid-1987, with the arrival of Bitnet. After eight months of connection, the network reached 110 nodes in January 1988, when the decision was taken to use Fapesp to implement an international connection to State universities.

Embratel, concerned with the monopoly it had in the area, only allowed the transport of third-party traffic through community networks research and academic research in October 1988, a month after the establishment of the first international connection. Thus, the first connection established with Bitnet, with a rate of 9600 baud, or symbols per second, was made between the LNCC, in Rio de Janeiro, and the University of Maryland, in the USA.

Then, negotiations began with the Central Committee of the Bitnet (Educom). The entry point was placed at Fermi National Laboratory (Fermilab), which had a Vax platform (Vax 750) and a cooperation contact signed with the Institute of Physics of the University of São Paulo (IFUSP).

In September 1988, Educom changed its rules, no longer accepting us at Bitnet in foreign countries. It was then necessary to structure a cooperation network, which changed the long-term plans. Negotiations finals were closed with Fermilab in October 1988 in Batavia, Illinois, and then with Educom. Thus, a cooperation network was established with the Bitnet, which became known as Academic Network at São Paulo (ANSP) (Alencar, 2008c).

The network's second international connection, which initially operated at a rate of 4800 baud, was installed in November 1988 between Fapesp and Fermilab in Chicago. This connection provided for the service of the university system and research center in São Paulo and used DECnet technology, which allowed access to *high energy physics network* (HEPnet) and Bitnet.

In February 1989, the Embratel line, from São Paulo, was installed in Batavia, USA, with HEPnet connection (DECnet) and address user@FPSP.HEPNET. It became the IFUSP DECnet node.

A third independent connection to Bitnet, also at 4800 bit/s, was installed in May 1989 between UFRJ and the University of California in Los Angeles (UCLA). The Federal University of Rio Grande do Sul (UFRGS) entered the network in July 1989, using the Phonenet protocol over X.25, a set of protocols standardized by the ITU for wide area networks, and using the telephone system or the integrated services digital network (ISDN) as a means of transmission.

In December 1989, a Vax6330 and an mVax3600 were purchased for Fapesp, with financial support from the Department of Science and Technology of the State, leaving the mVax3600 exclusively for network management. The international line has increased from 4800 to 9600 baud, in September 1990.

At the end of the 1980s, Embratel offered the following communication services: private lines, with transmission rates between 300 and 9600 bit/s; a switched data service called Renpac, with X.25 and X.28 access, at rates up to 9600 baud; and satellite service, using technology very small aperture transmitter (VSAT).

In May 1989, the Country had three islands of access to Bitnet. Two in the city of Rio de Janeiro, and one in São Paulo. Communications between these islands were through the international Bitnet network. The Federal University of Paraíba was interconnected to the network in 1989, with a direct connection with Embratel, in Recife.

After the elimination, by Embratel, of the restriction on traffic of third parties, it was possible to establish a national network to share the access to international networks. This was accomplished in 1991, with the interconnection between islands, and with the extension of connectivity to other research centers in the country.

The adoption of a final form of addressing by the National Network of Research (RNP), from CNPq, only took place in December 1990. The addresses in the form user@maquina.departamento.instucional.BR started to be used by institutions that had the capacity to manage their own sub-domain (Alencar, 2008d).

The addresses of the type user@maquina.departamento.instucional. ANxx.BR, remained for the other institutions. Pseudo-addresses related to Bitnet nodes, such as user@node.ANxx.BR, were formally eliminated, although their use has continued in practice.

Brazil's access to the Internet became possible in February 1991, when Fapesp increased the transmission rate to 9600 baud. Connection to Fermilab and installed the multinet program from TGV, to carry traffic that used the Internet Protocol (IP), in addition to DECnet, which was a proprietary network technology from Digital Equipment Corporation (DEC), and also Bitnet traffic.

IP connectivity has been extended to a small number of institutions in São Paulo, Rio de Janeiro, Rio Grande do Sul, and Minas Gerais, with low transmission rate private lines, between 2400 and 9600 bauds, or through Renpac.

The components of the second generation network were installed, in 1992, and included RNP and the state networks of Rio de Janeiro and São Paulo, which were financed by Faperj and Fapesp. State networks were installed for the United Nations Conference on Environment and Development (UNCED-92 or Rio-92), which was held in June 1992, in Rio de Janeiro, and both the networks used new international connections of 64 kbit/s. They served to support the Global Forum, a meeting of non-governmental organizations (NGOs), carried out in parallel to UNCED-92.

Bitnet was important for worldwide connectivity until the beginning of 1990s, when it was supplanted by the Internet. The main application of Bitnet was the maintenance of distribution lists. The most visible difference between Bitnet and the Internet was the addresses of the servers. The addresses of Bitnet had no points to separate server names from domains.

From the beginning of the 1990s, RNP started to provide access to approximately 600 teaching and research institutions, serving a wide range of community of about 65,000 users.

7.2 Layer Architecture

Computer networks use layered architectures, to organize the project and minimize its complexity. Each layer represents a logical entity that performs certain functions. The layers switch information through primitives, i.e., simple commands that define operations simple.

The services provided by the highest layer of the network, known as application, are sent directly to users. Each layer provides services for the next layer above it (Alencar, 2012a).

When users in different nodes communicate, corresponding layers also exchange information using rules that are appropriate. A set of rules, called a protocol, is necessary so that each level can communicate in a structured way. A layered architecture, with the indicated protocols, is shown in Figure 7.1.

Protocols are hierarchically organized, corresponding to the layers of the network. Protocols control exchange of information through a single layer and all the information layers communicate by logical or virtual paths. The lowest layer protocol is the only one to control the data flow in a physical connection channel.

The flow of information from one level to another is done through an interface that is present between each pair of adjacent layers. The purpose of the interface is to interpret formats and other characteristics of the protocols.

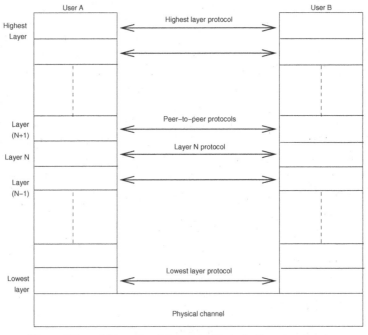

Figure 7.1   A layered architecture with the indicated protocols.

The number of layers, the name of each layer, and its function differ from one network to another. The most extreme layers are usually the physical and application layers. The application layer, as the name implies, handles user-specific applications such as access to files or graphic output from a computer assisted project. The physical layer is always the lowest level and is responsible for the transmission of signals between two nodes, providing an electrical or logical connection between processes.

When designing a network architecture, it must be taken into account that lower level protocols must be transparent to higher level protocols. Each layer must perform a defined function. The main advantages of layer architecture are flexibility to adapt lower level protocols without affecting those of the highest level and the ability to isolate well-defined functions (Alencar, 2012a).

Some major network architectures have become classic, including the IBM system network architecture (SNA) and the distributed network architecture (DNA) by Digital Equipment and the Department of Defense ARPANET from United States.

7.2.1 The ISO Reference Model

The International Standards Organization (ISO) developed a reference model for comparing of different architectures and construction of new networks. This model is called the open systems reference model for interconnection (OSI). The layers of the ISO reference model are illustrated in Figure 7.2 (Marksteiner et al., 2017), as compared to the TCP/IP model.

It is important to point out that the terminology for the system, in this context, can cover from one simple terminal to a complete computer network. The term "open" is used to show that the model applies to the transfer of information between interconnected systems, without concern with the internal operation of the systems themselves.

Generally, all layers are used at the origin and destination nodes, but only three are used in each intermediary node that only serves to transport packages to the destination nodes.

7.2.2 Reference Model layers

The layers of the ISO reference model are described in detail in this section. Briefly, the physical layer coordinates the necessary functions for the transmission information in a physical connection and the link layer

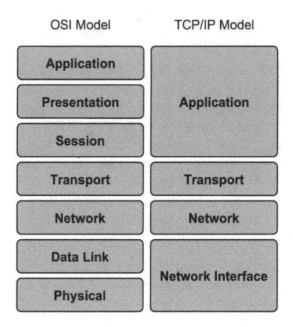

Figure 7.2   Layer structure of the ISO reference model compared to the TCP/IP model.

synchronizes the characters and messages and makes the transmission reliable.

For the transfer of packets between the source and the destination, there is the network layer; for this, the transport layer divides messages into packets, the session layer serves as an interface between the user and the network, the presentation layer simplifies communication between end users, and the application layer allows user access to the network.

Physical layer:

The electrical, mechanical, and functional characteristics, necessary for the transmission of the sequence of bits, are specified in the physical layer. These project specifications include signal features, such as amplitude, period, frequency spectrum (baseband or broadband, for example), and the modulation scheme that is used (Farouzan, 2008).

Details regarding physical connections, such as number of pins, as well as the mode of operation (full or half-duplex) are also considered. The physical layer synchronizes the bits, establishing, maintaining, and releasing the connection between the nodes.

Data link layer:

The data link layer provides the synchronizations of character and message (frame) and ensures reliable transmission blocks of data or frames between physically connected nodes. For this, the link layer creates and recognizes the limits (with the encapsulation) of the frames and uses redundant bits for the detection of errors.

The link layer retransmits frames with errors, producing error messages. Confirmations and managing other necessary details. The data link layer still controls the flow of packets to prevent the receiver from being overloaded and controls access, to determine which device takes control of the link at every moment.

Network layer:

The network layer performs the transfer of a data packet between the origin and the destination, typically through several links. It provides the services required to establish and maintain the flow of messages between users connected to the network.

The basic data unit of the network layer is the packet, and one of the functions of that layer is to ensure that packages are forwarded to their destinations. Considering that users are not necessarily connected by a direct physical link, this layer is responsible for routing and switching messages.

For forwarding messages, the network layer adds a header to the packet, which includes the logical addresses of the transmitter and receiver. The type of service provided to the transport layer by network layer is usually specified as datagram service or virtual circuit. In addition, the network layer controls flow and congestion on the network, preventing many packets from being routed to the same connection.

Transport layer:

The transport layer subdivides messages passed through the session layer into units (packages), if necessary, and allows these units to be transmitted to the intended destination. It performs segmentation of the message at the source and its assembly at the destination, identifying and replacing damaged packets during the transmission.

The transport layer also provides associated service application processes, which run on hosts or user devices. Its function is to transfer data transparently between application programs in the most independent of the possible network. Some of these services involve the delivery of e-mail messages, process by process, with appropriate ordering, the diffusion of messages for multiple destinations, and the delivery of isolated messages.

Session layer:

The session layer establishes and maintains a connection between application processes, serving as an interface for the user with the network. In addition, the layer can verify the authenticity of the user, provide charging, and decide the type of communication, for example, full-duplex or half-duplex.

It also allows for the exchange of information between two systems and makes it possible for the process to create verification in a data stream. There is no need to transmit the entire message in case of loss of part of the packets.

Presentation layer:

The presentation layer has the task of providing as many general functions as possible to simplify the communication between end users and protect the information that travels on the network. It is responsible for the syntax and semantics of the information exchanged between two systems.

The layer deals with the interoperability of different methods of encryption, encryption of information, to guarantee privacy, and data compression. Examples of services provided by the presentation layer are the conversion of codes, the text compression, and the use of layout standards for terminals and printers.

Application layer:

The application layer allows the user to access the computer network, such as the Internet, for example, and provides the right interface and support for certain services, such as message exchange, file transfer, establishment of a terminal virtual, which emulates a remote system terminal, and shared database management.

This layer has compositions and functions that depend on the user. Therefore, at this level, the software must contain the specific application programs, as a provision for the management of the required resources.

7.2.3 IEEE 802 Standards for Computer Networks

The Institute of Electrical and Electronic Engineers (IEEE) has committees that produce standards for all areas related to their professional activities, including those relat to computer networks, communication protocols, and modulation, among others.

The IEEE 802 standard, for example, defines the standards for local and metropolitan networks of physical and link layers of the OSI model for computer networks. The standards specify different types of networks, such as Ethernet, wireless network, and fiber optics, among others.

The IEEE 802 standards are divided into six basic parts, according to Figure 7.3, plus nine specific parts. The standards specify the physical and link layers of the ISO reference, with a unique logical link protocol and four types of media access technologies (Alencar, 2012a):

1. Standard IEEE 802.1 is a document describing network management, the relationship between the various parts of the pattern, and the relationship with the ISO reference model and with protocols for the highest level.
2. Standard IEEE 802.2 is a common protocol for the control of logical link (logical link control (LLC));
3. Standard IEEE 802.3 specifies the syntax and semantics for access control in the middle (media access control (MAC)). A bus that uses CSMA/CD as access method;
4. Standard IEEE 802.4 is a bus that uses token passing as access method (token bus (TB));
5. Standard IEEE 802.5 is a ring that uses token passing as access method (token ring (TR));

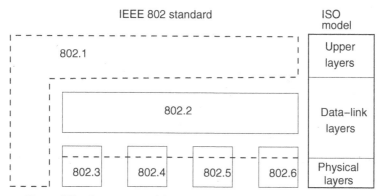

Figure 7.3 Components of the IEEE 802 standard and the ISO model.

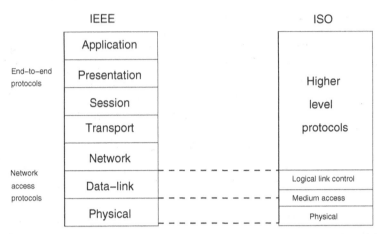

Figure 7.4   Comparison between the reference model for local IEEE networks and the ISO model.

6. Standard IEEE 802.6 is a metropolitan network (metropolitan access network (MAN));
7. Standard 802.7 defines broadband MAN;
8. Standard 802.8 defines the transmission in fiber optic network;
9. Standard 802.9 specifies the integration of local networks;
10. Standard 802.10 deals with security on local networks;
11. Standard 802.11 protocol for wireless local area network (LAN);
12. Standard 802.12 defines the demand-priority access method, physical layer, and repeater specifications;
13. Standard 802.15 defines the personal wireless network (wireless personal area network (WPAN)) (Bluetooth);
14. Standard 802.16 defines Broadband Wireless Access (BWA) (Wimax);
15. Standard IEEE 802.17 defines the resilient packet ring (RPR) access;
16. Standard 802.20 specifies mobile wireless access (WMA) (Mobile-fi);
17. Standard 802.22 describes the wireless regional area network (WRAN);
18. Standard 802.24 defined by the attributions of the Vertical Applications Technical Advisory Group that determines what enhancements to the horizontal technologies would improve their ability to support the applications.

The IEEE standard for local area networks presents, essentially, the same characteristics for transmission and reception of bits than the ISO model, as can be seen in Figure 7.4.

8

The TCP/IP Protocol

"If you think cryptography will solve your problem, either you don't understand cryptography, or you don't understand your problem."

Peter G. Neumann

8.1 TCP/IP Protocol History

In 1975, the Defense Advanced Research Projects Agency (DARPA), which originated as the ARPA, started to develop the TCP/IP protocol stack. In 1979, a committee was formed to lead the development of these protocols. This committee was called Internet Control and Configuration Board (ICCB).

DARPA has assigned the rights to the TCP/IP protocol code to the University of California for it to be distributed in UNIX version, in 1983, and requested that all computers connected to ARPANET use the TCP/IP protocols. Protocols spread quickly over the network.

Over the years, it has been realized that information can travel on communications networks in two different ways:

- Packet Switching – Where a line on the network can be shared by multiple information packages from different sources. Internet traffic is based on packet switching.
- Circuit switching – In which there is an exclusive line between the source and the destination. The transmission of information is done as a telephone call, in which a circuit is established between the parties throughout the communication.

8.2 Internet Coordination

The Internet is an articulated set of computer networks, which uses the TCP/IP protocol to packet communication, and contains many servers that

135

generally operate with operating systems UNIX and Linux. The Internet is coordinated by some entities, which take care of the organization of the network, distribution of IP addresses, creation of protocols, documentation, intermediation with government institutions, and the implementation of new technologies. The main bodies are:

- IAB – The Internet Advisory Board (IAB) is made up of several organizations and aims to coordinate the general organization of the Internet.
- InterNIC – The Internet Network Information Center (InterNIC) was created by NSF to distribute IP addresses.
- IRTF – The Internet Research Task Force (IRTF) is one of the committees that make up the IAB, which is responsible for research activities such as the development of protocols.
- RFC – The Requests for Comments (RFC) are technical documents related to Internet protocols. They can contain standards for the protocol, or can become standards. These documents form the Internet documentation.
- FNC – The Federal Networking Council (FNC) is a committee that produces information on the Internet. The FNC performs the intermediation between the IAB and government institutions, in addition to supporting agencies in the use of the Internet.
- IETF – The Internet Engineering Task Force (IETF) is a subcommittee of the IAB that deals with constructive problems of network and also with the implementation of new technologies.

The Internet works as follows. Suppose a machine on a local network wants to send a packet of information to a distant network. Initially, the packet is sent to the local network, where it is received by all machines and the router. The router checks the destination address of the packet, queries its route table, and sends it forward.

The packet travels through routers of various networks, in which it is sent to routers closer to the final address, until it reaches the destination machine. If the packet reaches the final address, the originating machine receives an acknowledgment message. If this message does not arrive within a certain time, the packet is retransmitted.

8.3 Types of Networks

The networks that make up the Internet can have several topologies, according to the coverage area, or coverage, and also depending on the technology used.

- LAN – Local area network is a local network located in a building or on campus. Its area is restricted to a few kilometers and its objective is to interconnect computers, workstations, and peripherals that share resources, such as printers.
- MAN – Metropolitan area network is a network that covers from some offices, neighborhoods, to an entire city.
- WAN – Wide area network is a network that covers a large geographic area, which can be a country or continent, usually formed by the interconnection of LAN networks.

8.4 Transmission Protocols

Protocols are the transmission agreements established between network servers to enable the communication. They can be connection oriented or not.

- Connection-oriented protocols – In this type of transmission, packets do not need to have overheads, as occurs in non-connection-oriented transmissions. At the beginning of the connection, the source and destination exchange all information needed for transmission. It is safer because it has mechanisms for forwarding packets with errors, as well as acknowledgment messages.
- Non-connection-oriented protocols – The origin and the destination need to have a prior agreement on the communication and on the characteristics of Quality of Service (QoS). This mode of transmission associates each packet with a global address, which identifies the source and destination of the package. Successive packages transferred are considered independent.

For non-connection-oriented transmission mode, flow control is not important, nor is there recognizing or resending packets. In this way, you can communicate with any machine without having to make a connection. On the other hand, there is no guarantee of the success of the transmission, and the monitoring of the process needs to be more effective.

8.5 Interconnection Equipment

Physical Internet connections rely on certain basic components:

- Router – A circuit or computer with routing software, which has ports connected to different networks. When he/she receives an information packet, he/she checks the destination address and looks for the equivalent port, based on network level addressing, which, in the case of TCP/IP architecture, is IP addressing. It tries to send the packet to networks that are closer to the destination network, decreasing the number of networks on which the packet travels.
- Gateway – It is a component of the network that aims to connect different networks, converting different levels of protocols, or in the case of IP, in order to perform the routing.
- Bridge – This device is used to connect similar networks. It forwards messages based on the level-two addressing, known as medium access control (MAC). It segments network traffic, creating a single collision domain if necessary.

8.6 Interconnection Protocols

The Transmission Control Protocol (TCP) performs communication between applications from two different hosts. TCP is a transport-level protocol that works with acknowledgment messages, specification of the information format, and security mechanisms. It ensures that all protocol data units (PDUs) are sent successfully, as it performs connection-oriented transmissions.

In addition, it allows the use of various applications aimed at conversation. When executed, it uses the IP protocol, not connection oriented. TCP is then responsible for control of secure data transfer procedures. For greater efficiency in communications, TCP encompasses several functions that could be in the applications, such as word processor, database, and electronic mail. It was created to be a universal software that contains these functions.

In addition, TCP performs additional services, such as the following:

- Flow control – Flow control assigns a transmission window to the source host. This window limits the number of bytes transmitted at a time.
- Transmission security – Reliability in transmissions via TCP in the connection orientation of the protocol and works with sequential and positive recognition numbers.

The TCP of the originating host transfers the data in the form of octets. Each octet is assigned numbers in sequence. The TCP of the target host analyzes these numbers to ensure the order and integrity of the message sent.

If the transfer is perfect, the TCP of the destination host sends an acknowledgment message to the origin. Otherwise, a numeric string is sent to the TCP from the host source that informs the type of the problem as well as order a new transmission.

Sequential numbers can also be used to eliminate duplicate octets, which can occur on account of non-connection-oriented transmission. The source TCP has a timer to ensure that you do not waste too much time between wrong messages and their correction. When the originating TCP receives an error message, a time-out occurs and the message is resent.

8.6.1 Other TCP Services

The TCP protocol performs other services, which include:

- OPEN/CLOSE commands – Through physical devices, TCP can establish a virtual connection, with the OPEN command. At that moment, TCP performs the three-way handshake, which is a process in which the source and destination TCP exchange acknowledgment messages to make the connection possible.

  By the time the transfer of information ends, any host (source or destination) can close the virtual connection with the CLOSE command.

- Information management in connection-oriented transmissions – The TCP protocol can control all aspects of the information being transmitted, as it is a connection-oriented transmission protocol.

- Priority and security – The TCP protocol allows the host administrator to control the levels of security and access permission, as well as connection priorities. These characteristics are not present in all versions, although they are defined in the TCP standards.

- Flow-oriented transfer stream – Interface applications send data to TCP in a targeted manner to stream. When the information arrives at TCP, it is grouped into packets and sent to the other transmission levels.

  The TCP protocol can use several ways of sending messages, including telephone lines, local networks, or high-speed fiber optic networks.

8.6.2 TCP Header Structure

The structure of the TCP header includes:

- Source port – Contains the source port number.
- Destination port – Contains the destination port number.
- Sequence number – The sequence number of the segment's first data octet (except when SYN is present). If SYN is present, the sequence number is the starting sequence number (ISN) and the first data octet is ISN + 1.
- Acknowledgment number – If the ACK control bit is activated, the field contains the value of the next sequence number that the recipient of the segment is waiting to receive.
- Data offset – Indicates where the data field starts within the TCP header.
- Reserved – 6 bits reserved for future use.
- Control bits – The control bit can be (from left to right):

    - U (URG) informs the application of the arrival of urgent data, which must be processed in the buffer.
    - A (ACK) indicates that the Acknowledgment field is important.
    - P (PSH) function Push.
    - R (RST) reinitializes the connection.
    - S (SYN) synchronizes sequence numbers.
    - F (FIN) indicates the end of data transmission.

- Window – 16 bits. The number of data octets that the recipient is waiting to receive, starting with the octet that indicates the Acknowledgment field.

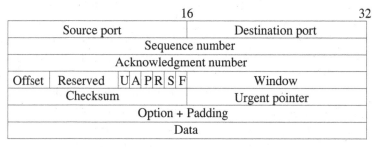

	16							32
Source port						Destination port		
Sequence number								
Acknowledgment number								
Offset	Reserved	U	A	P	R	S	F	Window
Checksum						Urgent pointer		
Option + Padding								
Data								

Figure 8.1   Structure of a TCP header.

8.7 The IP Protocol

The main function of the IP protocol is to transport datagrams from one network to another on the Internet. It is a non-connection-oriented transmission protocol, with the following characteristics:

- It has no retransmission mechanisms.
- There is no guarantee of full or orderly transmission.
- Uses IP addresses as a basis for targeting datagrams.
- Discards a datagram if it is not delivered or if you spend a lot of time traveling on the Internet.
- Its operations and standards are described in Request for Comments (RFCs) and internet Engineering notes (IENs).

Although the IP protocol does not have these characteristics, they are important. So data integrity transmitted remains with the TCP. Sending datagrams via IP goes through some basic steps.

8.7.1 IP Addresses

Each computer connected to the Internet has one or more IP numbers, each IP number on the Internet being unique, which prevents datagrams from being sent to the wrong place. A computer can play the role of host or gateway, or both.

IP addresses are made up of 4 bytes, separated by periods and divided into network address and local address.

The network address, provided by InterNIC, contains the part that refers to the main network and the subnets. It is represented in the first three bytes of the IP number.

The local address is assigned by the host administrator. It occupies the last byte and is used to identify local machines.

To facilitate the identification of addresses, the Domain Name System (DNS) was created, which associates a name to each IP number, with some characteristics:

- Names (domains) are separated by periods.
- There are no spaces between domains.
- The number of names can vary from one address to another, and they identify a single machine.
- There is no difference between uppercase and lowercase letters.
- Unlike the IP number, they are read from right to left.

8.7.2 Example of a DNS

DNS example: dee.ufcg.edu.br, in which:

- dee is the name of the machine (host);
- ufcg identifies the organization to which the host is connected;
- edu is part of the organization's identification;
- br is the field that represents the country (Brazil) or the type of the organization.

In addition to the common DNS hierarchy (country name, organization name, machine name), there may also be a special hierarchy, such as the names ufcg and edu, where edu indicates·an educational institution and ufcg is a university.

Associated with this address in the form of names can be the user's name (at the far left), separated by the symbol @ from the rest of the address, as in the example: malencar@dee.ufcg.edu.br, on what:

- malencar is the part that identifies the user (his login);
- @ is the character that separates the user's name from the rest of the address;
- iecom.org.br constitutes the remainder of the domain that goes from the name of the machine to the name of the country.

8.7.3 Datagram Fragmentation

When exchanging packages, Internet applications find a difference in size messages on different networks. The IP protocol allows for fragmentation, and datagrams are divided into smaller units.

The fragmentation procedure is performed by a gateway, and the messages are broken into units and identified. The destination host regroups the instructions based on the IDs of the gateway.

When identifying fragments, the gateway creates a header for each fragment, which contains the initial addresses of the networks (source address) and a message ID.

Upon regrouping, the destination host, upon receiving the first fragment, triggers a timer (Time to Live). If the standard time is exceeded and the message is not reconnected, the destination host discards fragments received and send an error message to the source.

8.7.4 IP Routing

A datagram sent by a host is not able to reach its destination without being guided by one or more routers.

A router receives the sent datagram and verifies its destination IP address, comparing it to its table. After this analysis, the datagram is sent to the gateway closest to the destination or even to the destination itself.

The table of a router can be static or dynamic. The dynamics are more flexible to adapt to the network modifications.

These tables determine the forwarding of datagrams to:

- Another subnet connected to the same router as the source network.
- A gateway, in case the network has access through one or more gateways (indirect routing).
- A default route if the address is not identified in the local table. In this case the usual default is to send the datagram to an external circuit.

Router tables are made up of four fields, as shown in Table 8.1.

- Field 1: IP address field. It contains all addresses sent by a given network.
- Field 2: This field contains control called metric, which uses algorithms to analyze the most efficient for a given address (including distance, equipment, and time).
- Field 3: The field that contains the IP address of the destination, be it a host from another subnet, a gateway, or an address default.
- Field 4: Field 4 contains the flags. A flag controls the frequency of use of a given address so that, in the case of pivot tables, the address can be deleted if it is little used.

Routers use a protocol to communicate or to communicate the information of status to hosts connected to them.

Table 8.1   Example of router tables.

143.54.0.0	134	143.55.0.1	1
200.17.162.00	122	143.54.0.3	0
200.132.0.0	59	143.44.0.2	1
200.111.0.0	172	143.54.1.22	0
Default	175	823.28.9.11	1

This protocol can be the Routing Information Protocol (RIP) or Open Shortest Path First (OSPF).

There is yet another protocol, the Internet Control Message Protocol (ICMP), which informs about possible errors in the routing process.

8.7.5 IP Address Resolution

Sometimes, when a host wants to send a message, he/she may not know the full address of the host destination, that is, to know only the IP number. For these cases, the IP protocol has a protocol called Address Resolution Protocol (ARP).

Two situations can occur in address resolution:

- The destination host is on the same network. Then, the originating host sends an ARP packet containing the destination IP number of the message to all other stations in your network. Only the station that identifies itself with the IP number of that packet will respond to the message, sending back your MAC address.

  In addition to sending the message, the originating host stores the MAC address received from the host from destination in a temporary table, which relates IP numbers to the MAC addresses of the supposed local network.

- The destination host is on a distant network. In this case, the originating host will send an ARP packet containing the IP number to the gateway, which then will guide this packet through several connections until it reaches the desired local network.

  Then there will be the same situation as in the previous one, with the difference that gateway itself will send ARP packets for all local stations.

The Reverse Address Resolution Protocol (RARP) is a protocol that allows a station to discover your own IP address. It is widely used by workstations without a hard disk since they cannot permanently store an IP address.

Each time a station of this type is going to send a message, it needs to discover its IP address. To do so, it sends a message to your local network's RARP server, containing its MAC address. The server returns its IP number to the station, thus enabling the sending of messages.

The ARP proxy is a variation of the ARP. It allows an organization to have only one IP address for its various networks.

In that case, all networks are connected to a router. When a host wants to communicate with a host from another network (without knowing its MAC address), it will dump a packet with the destination host IP number.

However, the packet is intercepted first by the router, which returns to the host destination MAC address itself.

Subsequent information is directed to the router, which re-writes it to the host destination, according to its own address table.

The structure of the header ARP/RARP is shown in Figure 8.2.

The structure of the ARP/RARP header is composed as follows:

- *Hardware type* – Specifies the type of hardware interface from which the issuer of the request expects a response from the request.

  Protocol type – Specifies the type of high-level protocol the issuer of the request has.

  HLen – Hardware address length.

  PLen – Protocol address length.

  Operation – Values are as follows:

  1 ARP request
  2 ARP response
  3 RARP request
  4 RARP response
  5 Dynamic RARP request
  6 Dynamic RARP reply
  7 Dynamic RARP error
  8 InARP request
  9 InARP reply

16		32
Hardware type		Protocol type
HLen(8)	PLen(8)	Operation
Sender Hardware Address		
Sender Protocol Address		
Target Hardware Address		
Target Protocol Address		

Figure 8.2 Structure of the ARP/RARP header.

The structure of the ARP/RARP header includes:

- Sender hardware address – Hardware address of the request sender.
- Sender protocol address – Address of the third-level (top level) protocol of the request sender.
- Target hardware address – Hardware address of the request recipient.
- Target protocol address – Recipient third-level (top level) protocol address.

8.7.6 User Datagram Protocol

The UDP protocol is restricted to ports and sockets, and transmits data in a non-connection-oriented manner. It is just an interface to the IP protocol.

The basic function of UDP is to serve as a multiplexer or demultiplexer for IP information traffic. Like TCP, it works with ports that properly guide the information traffic to each application. These ports are:

- Destination port – It is a part of the datagram (one end) that indicates the application to which whether to send the incoming information.
- Originating port – It is located at the other end of the datagram and indicates the application that sent the message. It can be used for a resend or, when not used, it is filled with zeros.

8.7.7 Applications that Use TCP and UDP

There are a number of top-level Internet applications such as e-mail that use TCP or UDP services. These applications are widely used Internet standards, mainly by its versatility.

TCP application protocols:

- TELNET – The telecommunications network protocol allows a user to work on a host distant. It emulates a special terminal that does the necessary conversions between two different terminals, allowing you to remotely act on a host without the need for both hosts to have a similar terminal.
- FTP – The File Transfer Protocol is a tool for transmitting files over the Internet. It defines the procedures for managing the exchange of information between TCP hosts. An FTP connection goes through two processes: control connection and data transfer.

The control connection is the first step in the FTP connection process. It serves to hold the host and define security and file manipulation levels. Data transfer is the stage at which files are transmitted. It depends on the success of control connection to be made.

8.7.8 Applications that Use TCP and UDP

The TCP application protocols are:

- SMTP – Simple Mail Transfer Protocol is a top-level application that is connected to the transmission e-mail via the Internet.
  It is one of the mostly used top-level protocols on the Internet, which works as follows. SMTP is made up of two parts, which are the source and the destination, each of which has access to a storage server.
  When the source sends a message to the destination, this message is stored on the source's storage server. The server then tries to send messages and, if there is a problem with the destination, the server will later try to resend the message. If not, the message will be sent back to the source or to postmaster.
- SNMP – The Simple Network Management Protocol is the most expressive standard in terms of network management. It is a protocol that is used to deal with eventual network or equipment failures. SNMP is still used for monitoring networks, mainly in networks that use TCP/IP.
- RPC – Remote Procedure Call is a protocol that allows a host to use a function located on a remote host. RPC allows the exchange of messages, in which the source sends parameters to a server and is waiting for a return, which provides the result of the remote function.
- TFTP – Trivial File Transfer Protocol is one of the most elementary of all, as it has few utilities, such as time and integrity control. It is a simpler model of FTP, as it has no mechanism security, and as it is based on the UDP protocol, the integrity of your transmissions cannot be trusted. This protocol is not widely used today, but some vendors still distribute it to their customers. equipment to avoid possible incompatibilities.

8.8 The TCP/IP Protocol

TCP/IP is the most widely used network protocol today. A protocol is a set of rules to allow two or more computers to communicate.

TCP/IP is a set, or a stack of protocols, as it is known. The acronym refers to two different protocols, TCP (Transmission Control Protocol) and the IP (Internet Protocol).

There are many other protocols that make up the TCP/IP stack, such as FTP, HTTP, SMTP, and UDP.

TCP/IP has four layers. The programs communicate with the application layer. In the application layer, you can find application protocols such as SMTP (for e-mail), FTP (for file transfer), and HTTP (for network browsing).

After processing the program request, the protocol in the application layer communicates with another protocol in the transport layer, usually TCP.

This layer is responsible for taking the data sent by the upper layer, dividing them into packages, and sending them to the immediately lower layer, the Internet layer. In addition, during data reception, this layer is responsible for placing packets received from the network in order (since they may arrive out of order) and also check that the contents of the packages are intact.

In the Internet layer, there is the IP, which takes the packets received from the transport layer and adds information address, or IP address (the address of the computer sending the data and the address of the computer receiving the data).

Then the packets are sent to the next lower layer, the Network Interface layer. In this layer, the packets are called datagrams.

The network interface layer receives the packets sent by the Internet layer and sends them to the network (or receives data from the network, if the computer is receiving data). What is in that layer depends on the type of network the computer is connected to.

Currently, computers use the Ethernet network and, therefore, the layers of the network interface must be found in the Ethernet, which are logical link control (LLC), media access control (MAC), and physics. Packets transmitted over the network are called frames.

8.8.1 Application Layer

This layer communicates between programs and transport protocols. There are several protocols that operate at the application layer. The most known are:

- HTTP (HyperText Transfer Protocol);

- SMTP (Simple Mail Transfer Protocol Mail Transfer);
- FTP (File Transfer Protocol);
- SNMP (Simple Network Management Protocol;
- DNS (Domain Name System);
- Telnet, a protocol for communication between remote terminals.

When an e-mail client program wants to download e-mails that are stored on the e-mail server, it makes this request to the application layer of TCP/IP, being served by the SMTP protocol.

When you enter a www address to view a page on the Internet, it communicates with the layer application of TCP/IP, being served by the HTTP protocol.

The application layer communicates with the transport layer through a door. The doors are numbered and the doors standard applications always use the same ports.

For example, the SMTP protocol always uses port 25, the HTTP protocol always uses port 80, and FTP uses ports 20 (for data transmission) and 21 (for transmission of control information).

The use of a port number allows the transport protocol (TCP) to know what type of content of the data package (for example, if it is e-mail), and receiver, to know which application protocol it should deliver the data packet to.

Upon receiving a packet destined for port 25, the TCP protocol delivers it to the protocol that is connected to this port, typically SMTP, which in turn delivers the data to the application that requested it (the e-mail program).

Figure 8.3 illustrates how the application layer works.

8.8.2 Transport Layer

In data transmission, the transport layer is responsible for taking the data passed through the data layer and application layer and turn them into packages.

TCP is the mostly used protocol in the transport layer. At the reception of data, the TCP protocol takes the packets passed through the Internet layer and puts them in order since the packets can reach the destination out of order, check if the data inside the packets are intact, and send a confirmation signal called acknowledge (ACK) to the transmitter, informing that the packet has been received correctly and that the data is healthy.

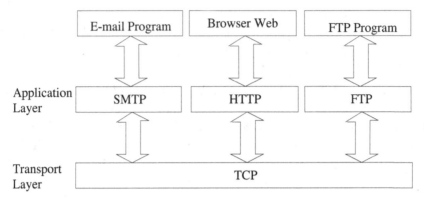

Figure 8.3   How the application layer works.

If no acknowledgment signal (ACK) is received (either because the data did not reach its destination or because TCP discovered that the data was corrupted), the transmitter will resend the lost packet.

While TCP reorders packets and uses an acknowledgment mechanism, which is desirable in data transmission, there is another protocol that operates in this layer that does not have these resources. This protocol is UDP (User Datagram Protocol).

For this reason, TCP is considered a reliable protocol, while UDP is considered an unreliable protocol. UDP is typically used when no important data is being transmitted, such as DNS requests. Since UDP does not reorder packets or use an acknowledgment mechanism, it is faster than TCP.

When UDP is used, the application that requests the transmission is responsible for verifying that the data received are intact or not and also to reorder the received packets, that is, the application does the work of TCP.

During data transmission, both UDP and TCP receive data passed from the application layer and add a header to that data.

Upon receipt of data, the header will be removed before the data is sent to the appropriate port.

This header contains control information, in particular, the number of the source port, destination port number, sequence number (for acknowledgment of receipt and reordering mechanisms used by TCP), and a checksum (called checksum or CRC, which is a calculation used to verify that the data was received intact at the destination).

The UDP header is 8 bytes, while the TCP header is between 20 and 24 bytes, depending on whether the options field is being used or not.

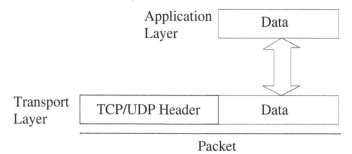

Figure 8.4 Data packet generated at the transport layer.

Figure 8.4 illustrates the data packet generated at the transport layer. This data packet is sent to the Internet layer, for data transmission, or is received from the Internet layer, when receiving data.

8.8.3 Internet Layer

In TCP/IP networks, each computer is identified with a unique virtual address, called an IP address. The Internet layer is responsible for adding a header to the data packet received from the transport layer where, among other control data, the source IP address and the destination IP address are added, that is, the IP address of the computer sending the data and the IP address of the computer that should receive it.

Each computer's network card has a physical address. This address is recorded in the card's ROM memory network and is called MAC address. That way, on a local network, if computer A wants to send data to computer B, it needs to know the MAC address of computer B. In a small local network, it is possible for the computers to easily find out the MAC address of all PCs, but the task is not simple on a global network like the Internet.

If no virtual addressing scheme is used, you need to know the MAC address of the target computer, which is not only a complicated task but also does not help in the routing of packets, as this address does not use a tree structure.

Routing is the path that the data must use to reach the destination. When someone requests data from an Internet server, for example, this data goes through several locations (called routers) before it arrives the computer.

In all networks connected to the Internet, there is a device called a router (which bridges between computers on the local network and the Internet). Every router has a table containing the known networks and also a configuration called the default *gateway* pointing to another router on the Internet.

When the computer sends a data packet to the Internet, the router connected to the network first checks if it knows the destination computer; in other words, the router checks whether the destination computer is located on the same network or on a network that knows the route.

If it does not know the route to the destination computer, it will send the packet to its default gateway, which is another router. This process is repeated until the data packet reaches its destination.

There are several protocols that operate at the Internet layer:

- IP (Internet Protocol);
- ICMP (Internet Control Message Protocol);
- ARP (Address Resolution Protocol);
- RARP (Reverse Address Resolution Protocol).

Data packets are sent using the IP protocol. The IP takes the data packets received from the transport layer and divides them into datagrams.

The datagram is a package that does not contain any type of acknowledgment of receipt (ACK), which means that IP does not implement any receipt acknowledgment mechanism, that is, it is an untrusted protocol.

It should be noted that during data transfer, the TCP protocol will be used over the Internet layer (that is, over IP) and TCP implements the receipt acknowledgment mechanism.

Therefore, although the IP protocol does not verify that the datagram has arrived at the destination, the TCP protocol will do this verification. The connection will then be reliable, although IP alone is an unreliable protocol.

Figure 8.5 illustrates a datagram in the Internet layer.

Each IP datagram can have a maximum size of 65,535 bytes, including its header, which can use 20 or 24 bytes, depending on whether a field called options is used or not. That way IP datagrams can carry up to 65,515 or 65,511 bytes of data.

If the data packet received from the transport layer is larger than 65,515 or 65,511 bytes, the IP protocol will fragment packets into as many datagrams as needed

The header added by the IP protocol includes the source IP address, the destination IP address, and several other control information.

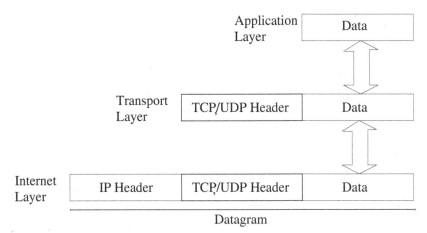

Figure 8.5 Datagram in the Internet layer.

8.9 Security Protocols for TCP/IP

The protocol architecture that governs data transmission on computer networks that make up the Internet, designated by the acronym TCP/IP, initially foresaw an open network environment with few security mechanisms. This provided the public nature of the references (Request For Comments – RFC) and standard implementations of the TCP/IP protocol suite.

This public and open nature, without protection mechanisms, made security is one of the main structural problems of the TCP/IP architecture. This opening was responsible for the success and popularization of. However, the deficiency in the security aspect represents one of the limitations to the use of technology, especially in commercial applications.

Among the deficiencies, in the security aspect, of the IP protocol is its inability to authenticate a machine on the network. Based on the source IP address of a received packet, it is not possible to determine with certainty the identity of the machine that has it originated. There is also no guarantee that the contents of a received package are unchanged or that data privacy has been preserved (Pouw, 1999).

The initial solution to provide network security was to implement individual algorithms, for each application, when the needs arose. In this way, the PGP was created to communication by e-mail, SSL/TLS for Internet browsing, and SSH for login secure remote. However, although implementation at the application layer is simpler since it does not involve the operating system and is restricted to a specific application, a service implemented in this layer makes it specific, creating the need for a new development for each new application.

The most appropriate solution was to implement the security service in the inter-network layer, which delivers packets without connection, which allows control by flow or connection. Some protocols were developed for this layer for this purpose, some run only on IP (as in the case of GRE and PPTP), and others, more versatile, are able to handle not only IP packets but also IPX and NetBEUI (as in the case of L2F or L2TP).

The guarantee of security on the Internet requires that the IP protocol itself offers this service, without relying on other protocols. The integration of this functionality with the new version of IP (IPv6) was proposed and its development was in charge of the working group IP Security Protocol (IPSec) the Internet Engineering Task Force (IETF).

Considering the delay in the migration of the Internet to the new version of IP, several adaptations were made for IPSec to run over IPv4. Its implementation was made with special headers, with header extensions in IPv6 and an additional protocol header in IPv4, typically placed between the original IP header and its payload.

8.9.1 IP Security Protocol

The IP Security Protocol is an open platform formed by a set of protocols that provide the following security services:

- access control;
- package integrity;
- origin authentication;
- package privacy;
- privacy in packet flow;
- protection against replays.

Having an open architecture makes it possible to include other authentication and encryption algorithms, in addition to the necessary HMAC-MD5 and HMAC-SHA-1, for authentication and encryption, and DES-CBC, for encryption only. This topic will be detailed in other chapters.

8.9.2 Vulnerabilities and Precautions Regarding TCP/IP

Vulnerabilities in the TCP/IP protocol stack are exploited through the following attacks.

SYN Flooding – It consists of sending several packets to open a data exchange connection (SYN packets), containing a forged IP address of a machine that is not operational and, therefore, that is not able to complete the connection opening process.

IP Spoofing – When the attacker masks your computer as a legitimate network machine. This is done by forging the source address of IP datagrams. Thus, the attacker associates his or her datagrams with a legitimate IP address of the target network.

Sequence Number Attack – The TCP protocol uses a sequence of numbers associated with the packets to ensure that the packets arrive at the destination in the order in which they were sent. The TCP sequence number prediction attack consists of predicting the next sequence number of the packet that travels and thus forging packets with the numbering expected by the target machine.

TCP Session Hijacking – This attack is a variation of the attack by sequence number, which also uses a forged IP address. This attack can be performed against any application based on the TCP protocol, such as Telnet, rlogin, and FTP. It consists of hijacking the connection between two machines: a client machine and a server.

RST and FIN attacks – The TCP protocol has flags that are used to control the connection. An attacker can use the RST and FIN flags to generate a denial of service attack. Normally, the RST flag is used to restart a connection, and the FIN flag is used to indicate that there is no more data to be sent.

Ping O'Death – The Ping program is used to test whether a machine is operational, by sending a packet of the type "ICMP echo" and waiting for the receipt of this packet to be received by the target machine. Older versions of this program make it possible to send a package with a larger size than allowed. When the target machine receives this packet and starts processing, system variables overflow may occur.

Possible measures against these attacks on the TCP/IP stack are:

- block the sending of packets whose source address does not belong to the set of addresses that make up the local network;
- block the receipt of packets that originate on the Internet and that contain a source address equal to an IP address on the local network;
- use encryption and authentication in communication between machines via the Internet and in all TCP services;
- periodically update the operating systems (Ping O'Death).

8.9.3 Vulnerabilities and Precautions Regarding TCP/IP Services

Service Vulnerabilities and Precautions:

The vulnerabilities of the services offered by TCP/IP are associated with certain stack protocols:

Telnet – This is a program used for communication between hosts. The vulnerability of Telnet is that it does not protect the user's login and password during transmission to the remote machine.

Finger – It is used to obtain information about the users of a given host, such as the user's name and the last time he had access to the system. An attacker could use information provided by the finger command to perform an attack, such as, for example, using the user's login and name to try to find out their password.

File Transfer Protocol (FTP) – FTP is vulnerable because, when establishing a connection between a client and an FTP server, the user's login and password are sent without protection, and an attacker who is monitoring packet traffic on the network can discover the user's password.

The usual solutions to these problems are listed below:

- periodically update the network's software;
- limit the services offered by the network's servers;
- inspect the system's logs files;
- check running processes and strange files;
- validate system settings;
- educate users about network security;
- limit users' access to the system;

- disable the Telnet and Finger commands;
- run frequent backups;
- constant updation of administrators and those responsible for security in relation to method system invasion;
- encourage the use of cryptography on the network;
- install firewalls to filter incoming packets.

9

Network Security

"One must acknowledge with cryptography no amount of violence will ever solve a math problem."

Jacob Appelbaum, Cypherpunks: Freedom and the Future of the Internet

9.1 Cryptography Applied to Computer Networks

The Internet has revolutionized the way companies do business, given that the Internet Protocol (IP) is undeniably efficient, inexpensive, and flexible (Alencar, 2011a, 2012a). However, poorly configured networks are vulnerable to attack by intruders. Therefore, it is necessary to guarantee the company's business by protecting what is most important to its operation, the information. For this, it is necessary to respect the fundamentals of information security (Rocha Jr., 2013):

- Integrity – Certifies that the information is not modified, between source and destination, during communication.
- Confidentiality – Restricts information only to parties authorized by network management.
- Authenticity – Ensures that the information comes from an legitimate source.
- Availability – Ensures the provision of the service to a legitimate user (Lima et al., 2014).
- Non-repudiation – The set of measures to prevent the sender or recipient to deny the transmitted message.
- Responsibility – The characteristic of the system that allows to provide audit trails for all transactions.
- Access control – Access is allowed only to authorized users, through identification and authorization.

159

This section presents the main aspects of modern TCP/IP computer network encryption, including digital signature technology based on asymmetric encryption algorithms, data confidentiality that applies symmetric cryptographic systems, and public key infrastructure (PKI) (Tanenbaum, 2003).

The potential vulnerabilities of TCP/IP networks, and techniques for eliminating them, are considered. Note that only a general, multi-layer security infrastructure can handle possible attacks to computer network systems.

9.1.1 Potential Network Vulnerabilities

The Internet Protocol (IP) is efficient, inexpensive, and flexible. However, the existing methods used to route IP packets leave them vulnerable to security risks, such as spoofing, sniffing, and session hijacking, and provide no form of non-repudiation for contractual or monetary transactions.

Organizations need to provide security for communications between remote offices, business partners, customers, and traveling and telecommuting employees, besides securing the internal environment. The transmission of messages over the Internet or intranet poses a risk, given the lack of protection at the existing Internet backbone.

Control and management of security and access between the entities in a company's business environment is important. Without security, both public and private networks are susceptible to unauthorized monitoring and access. Internal attacks might be a result of minimal or non-existent intranet security.

The usual risks from outside the private network originate from connections to the Internet and intranets. Password-based user access controls alone do not protect data transmitted across a network. The data might be subjected to attack without the installation of security measures and controls. Some attacks are passive, and the information is only monitored. Other attacks are active, and the information is altered to corrupt or destroy the data or the network itself.

9.1.2 Listening, Data Change, Forged Identity

Most network communications occur in a plaintext, or non-encrypted, format, which allows an attacker, with access to data paths in a network, to monitor and interpret the traffic. When an attacker is tapping, or eavesdropping, to communications, it is referred to as sniffing or snooping.

The ability of a spy to monitor the network is the biggest security problem that administrators face in a company. Without good encryption services that are based on cryptography, data can be read by others when it is transmitted over the network. After an attacker has read the information, the next logical step is to modify it. An attacker can modify the data in a packet, without the knowledge of the sender or receiver.

Most networks and operating systems use the IP address to identify a computer as valid in a network. However, it is possible for an IP address to be misused, which is known as identity spoofing. An attacker uses special programs to build IP packets that appear to have originated from a valid address inside an organization intranet. After gaining access to the network with a valid IP address, the attacker can modify, redirect, or delete data.

9.1.3 Password-Based Attacks

It is common for most operating systems, and network security plans, to use password-based access control. Access to both a computer and network resources is granted using a username and password. Older versions of operating system components did not always protect identity information when it passed over the network to be validated. This can allow a spy to obtain a valid username and password and use them to gain access to the network by posing as a valid user. (Markovic, 2002).

When an attacker finds and accesses a valid user account, he/she is given the same rights as the real user. For example, if the user has administrator rights, the attacker can create additional accounts to access later. After gaining access to a network with a valid account, an attacker can obtain lists of other users and computer names, as well as network information, and can also modify server and network settings, including access controls and routing tables, being able to modify, redirect, or delete data (Markovic, 2002).

9.1.4 Denial of Service

Different from password-based attacks, the denial-of-service (DoS) attack prevents normal use of a computer or network by valid users. Unlike most attacks on the Internet, denial-of-service is not intended to acquire confidential information, alter information, or disseminate malicious programs on the network. Its purpose is to make specific computer network services unavailable to legitimate users.

This technique consists of sending a large amount of IP packets, typically on TCP, UDP, or ICMP protocols and, more recently, HTTP. In addition to taking up resources on the target computer, including processing, memory, limit connections, or conversations available, these attacks usually occupy a large part of the network bandwidth in which the attack recipient is located, often reaching the capacity of the network and making communication impossible (Queiroz et al., 2014).

After gaining access to a network, via some strategy, an attacker can:

- Distract the management of the information system, so that the intrusion is not immediately noticed. This gives the attacker the opportunity to make other attacks.
- Send invalid data to applications or network services, causing applications and services to close or operate abnormally.
- Send torrent of traffic until the computer or an entire network is brought down.
- Block traffic, which results in loss of access to network resources by authorized users.

9.1.5 Password Break Attack

As discussed in the previous chapter, a password is a code, or number, needed to encrypt, decrypt, or validate protected information. It is possible for an attacker to obtain a key, although it is a difficult and resource-intensive process. After an attacker obtains a key, it is called a corrupted, or compromised, key.

An attacker uses a corrupted key to gain access to protected communications, without the sender or recipient being aware of the attack. With the compromised key, the attacker can decrypt or modify data. The attacker can also try to use the corrupted key to compute other keys, which can allow access to other protected information.

9.1.6 Sniffer Attack

A sniffer is an application or device that can read, monitor, and capture exchanges of data and packets on the network. If the packages are not encrypted, a sniffer provides a complete view of the data that is in the packet. Even encapsulated packets can be opened and read if they are not encrypted.

Using a sniffer, an attacker can scan a network, analyze its contents, and access information, eventually causing the network to stop responding or become corrupted, and also read private communications.

9.1.7 Attack to the Application Layer

An application layer attack is directed at application servers to cause a failure in a server's operating system or applications. After the attack has taken place, the attacker is able to bypass normal access controls.

The attacker takes advantage of this situation, gaining control of an application, system, or network, in order to read, add, delete, or modify data or an operating system and introduce a virus that uses software applications to copy viruses across the network.

The attacker can also introduce a sniffing program to analyze the network and obtain information that can eventually be used to cause the network to stop responding or become corrupted, to close data applications or operating systems abnormally, and to disable other security controls to allow future attacks. Virus attacks are the main concern for computer users, accounting for 94% of all error reports sent, followed by internal abuse of network access, with a percentage of 91% (Markovic, 2002).

9.2 Alternatives to Prevent Attacks

Possible ways to prevent attacks are listed below (van der Lubbe, 1997):

- Encryption, which means protecting data and password confidentiality.
- Use of digital signature technology to ensure authenticity, integrity protection, and non-repudiation.
- A strong authentication procedure to allow secure authentication between communicating parties.
- Use of strong keys and frequent change of keys to combat cryptanalysis methods.
- Network address translation to protect against denial of service attacks.
- Use of PKI digital certificates as a unique electronic identification between communicating parties.
- Use of smart cards to generate and store secure key as well as generate digital signatures.
- Use of anti-virus, anti-spam, and anti-phishing protections.
- Use of intrusion prevention systems.

It is important to remember that remote transactions require privacy to preserve the confidentiality of the parties. Authentication is important to compensate for the lack of visual or physical contact during communication and to prevent identity spoofing. Remote transactions also require integrity to prevent data corruption and non–repudiation to prevent false transactions.

9.2.1 Security Technologies

The main cryptographic aspects of modern TCP/IP networks are: digital signature technology based on asymmetric encryption systems, confidentiality protection based on symmetric cryptographic algorithms, and public key infrastructure (PKI) (Forouzan and Mosharraf, 2011).

To avoid malicious attacks on the network, a multi-layer security architecture must be implemented. Modern computer network security systems consist of security mechanisms in three different layers of the ISO/OSI reference model:

- Application-level security, or full security (end-to-end), based on strong user authentication, digital signature, confidentiality protection, digital certificates, and hardware tokens, such as smart cards.
- Transport-level security based on the creation of a cryptographic tunnel, or symmetric cryptography, between network nodes, and a robust node authentication procedure.
- Security at the IP network level that guarantees protection against attacks from external networks.

These layers are designed so that a vulnerability in one layer does not compromise the others; so the entire system does not become vulnerable.

9.2.2 Security Mechanisms for the Application Layer

Security mechanisms at the application level are based on symmetric and asymmetric encryption systems, which perform the following functions:

- Authenticity of the due parts, for asymmetric systems.
- Protection of the integrity of the transmitted data, for asymmetric systems.
- Non-repudiation, for asymmetric systems.
- Protection of confidentiality at the application level, for symmetrical systems.

The currently mostly used protocols in the application layer domain are: S/MIME, PGP, Kerberos, application-level proxy servers, crypto APIs, and SET for applications of the type client-server. Most protocols are based on X.509 PKI certificates, a digital signature based on asymmetric algorithms, such as RSA, and confidentiality protection based on symmetric algorithms such as DES, 3DES, IDEA, and AES.

Most modern application-level protocols, such as S/MIME and crypto APIs, in client-server applications, are based on digital signature and digital envelope technologies. Security systems, at the application level, also have an authentication procedure that can rely on three components.

9.2.3 Security Mechanisms for the Transport Layer

Security mechanisms in the transport layer include the protection of confidentiality of transmitted data based on symmetric encryption algorithms. These systems are, for the most part, based on the establishment of the cryptographic tunnel between two nodes of the network, at the transport level, which is preceded by a strong authentication procedure.

In this way, the systems are based both on symmetric algorithms, to perform the cryptographic tunneling, as well as a bilateral challenge–response authentication procedures, based on asymmetric algorithms and digital certificates for authentication of nodes and for the establishment of a symmetric session key for the tunneling session.

The transport-level security system is mainly used to protect communication between the client and the Internet browser, such as Chrome, Safari, Firefox, Internet Explorer, Netscape Navigator, and WEB servers. And the most popular protocols are: SOCKS, SSL/TLS, and WTLS. Among them, the most popular, and by far the most used, is the secure sockets layers (SSL) protocol layer, which is used for protection between the client's browser program and the WEB server.

9.2.4 Security Mechanisms for the Network Layer

Security mechanisms for the network level include mechanisms implemented in communication devices, firewalls, and security mechanisms of the operating system. Using these methods, it is possible to implement virtual private networks (VPN). Security protection is achieved through complete encryption of IP traffic, using link encryption between two nodes on the network.

The most popular network layer security protocols are: IPSec (AH and ESP) and packet filtering and network tunneling protocols. IPSec is the most widely used. IPSec, as well as other transport-level security protocols, consists of authentication of network nodes based on asymmetric encryption algorithms and link encryption based on symmetric algorithms.

Firewalls, which can be computers, routers, and workstations, have, as their main characteristic, the definition of which information and services on the internal network can be accessed by the external world, and which internal users are allowed to use external information.

Firewalls are installed mainly at breakpoints that connect insecure external networks and the secure internal network. Depending on the needs, firewalls consist of the one or more functional components from the following set: packet filter, application level gateway, and circuit level gateway. There are four important examples of firewalls:

- packet filtering firewall;
- dual-homed firewall, with two network interfaces, each communicating with the corresponding network (internal and external);
- screened host firewall, with a router designed to filter service packets on the network;
- Screened subnet firewall, with a secure subnet between the internal and external networks, called DeMilitarized Zone (DMZ).

9.3 Secure Sockets Layer Protocol

Originally developed by Netscape, the SSL protocol has been universally accepted, for authenticated and encrypted communications between clients and servers. The new standard of the Internet Engineering Task Force (IETF), called Transport Layer Security (TLS), is based on SSL. This was recently published as an ongoing IETF document, titled TLS Protocol Version 1.0.

The Transmission Control Protocol/Internet Protocol (TCP/IP) commands the transport and routing of data over the Internet. Other protocols, such as the HyperText Transport Protocol (HTTP), Light Directory Access Protocol (LDAP), or the Internet Message Access Protocol (IMAP) work above TCP/IP, in the sense that they use TCP/IP to support typical application tasks such as displaying web pages or running e-mail services (Tanenbaum, 2003).

The SSL protocol works above TCP/IP and below higher level protocols, such as HTTP and IMAP. It uses TCP/IP by the highest level protocols, and

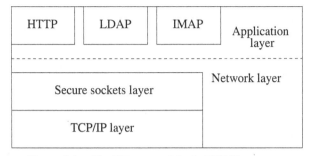

Figure 9.1   The SSL protocol in the TCP/IP structure.

in the process, it allows an SSL-enabled server to authenticate to an SSL-enabled client, and vice versa, as well as allows both machines to establish an encrypted connection. In Figure 9.1, one can see the SSL protocol in the TCP/IP structure.

The capabilities address fundamental concerns about communication over the Internet and other TCP/IP networks:

- SSL server authentication allows the user to confirm the identity of the server.
- SSL client authentication allows the server to confirm the client's identity.
- An encrypted SSL connection requires that all information transmitted between client and server be encrypted by the transmitting software and decrypted by the receiving software, thus providing a high degree of confidentiality.

The SSL protocol includes two sub-protocols: the SSL recording protocol and the handshake SSL protocol. The SSL recording protocol defines the format used for data transmission. The handshake protocol involves using the SSL recording protocol to exchange a set of messages between an SSL-enabled server and an SSL-enabled client when they establish a first SSL connection.

This exchange of messages is designed to facilitate the following actions:

- authenticate the server to the client;
- allow client and server to select the cryptographic algorithms, or ciphers, that both support;
- optionally authenticate the client to the server;
- use public-key cryptography techniques to generate shared secrets;
- establish an encrypted SSL connection.

9.3.1 Encryption Used with SSL

The SSL protocol supports the use of a variety of cryptographic algorithms, or ciphers, for use in operations such as server and client authentication, transmission of certificates, and establishment of session keys.

Clients and servers can support different packages, or sets, of ciphers, depending on factors such as which version of SSL is supported, company policies related to acceptable cryptographic strength, and government restrictions on the export of enabled software with SSL.

Among other functions, the handshake SSL protocol determines how the server and the client negotiate which encryption packages they will use to authenticate themselves, transmit certificates, and establish session keys. The usual figures are:

- data encryption standard (DES), a cryptographic algorithm used by the United States government;
- digital signature standard algorithm (DSA), part of the digital authentication standard used by the United States government;
- key exchange algorithm (KEA), an algorithm used to exchange keys by the United States government;
- message digest (MD5), an algorithm developed by Rivest;
- Rivest cryptographic figures (RC2 and RC4), developed for RSA data security;
- A public key algorithm for both encryption and authentication, developed by Rivest, Shamir, and Adleman (RSA);
- a key exchange algorithm for SSL based on the RSA key exchange algorithm;
- secure hash algorithm version 1 (SHA-1), a function of hash (univocal function that transforms any data line into a fixed-length line) used by the United States government;
- a secret symmetric key algorithm, that is implemented in hardware, known as FORTEZZA, used by the United States government (SKIPJACK);
- DES applied three times, or triple DES.

Key exchange algorithms, such as RSA and KEA key exchange, determine how the server and client will define the symmetric keys that they will both use during an SSL session. The most common SSL encryption packages use RSA key exchange.

The SSL 2.0 and SSL 3.0 protocols support overlapping cipher suites. Administrators can enable or disable any of the cipher suites for both clients and servers. When a given client and server exchange information during SSL handshake, they identify the strongest encrypted packet they have in common and use it for the SSL session.

Decisions about which encryption packages a given organization decides to enable depend on a balance between the sensitivity of the data involved, the speed of the encryption, and the applicability of export rules. Some organizations may want to disable weaker ciphers to prevent SSL connections with weaker encryption.

To serve the widest possible range of users, it is important that administrators enable a wide variety of SSL encryption packages. So when a client or home server deals with another server or home client, respectively, it negotiates the use of the strongest available cipher.

When a domestic client or server deals with an international server or client, it negotiates the use of those ciphers that are permitted under US export regulations. However, since 40-bit ciphers can be broken relatively easily, administrators who are concerned about eavesdropping and whose user communities can legally use stronger ciphers should disable the 40-bit ciphers.

9.4 Exchange of Information for the Safe Sockets Layer

The SSL protocol uses a combination of public key and symmetric key cryptography. Symmetric key encryption is faster than public encryption, but public encryption allows for better authentication techniques.

An SSL session always begins with an exchange of messages called the SSL handshake. The handshake allows the server to authenticate itself to the client using public key techniques and then allows the client and the server to cooperate in the creation of symmetric keys used for fast encryption, decryption, and tamper detection during the following session. Optionally, the handshake also allows the client to authenticate itself to the server.

The client sends the SSL version number, encryption parameters, randomly generated data, and other information that the server needs to communicate with the client using SSL to the server.

The server sends the client its SSL version number, encryption parameters, randomly generated data, and other information that the client needs to communicate with the server using SSL. The server also sends its own certificates, and if the client wants a server resource that needs authentication from the client, it requests the client's certificate.

The client uses some of the information sent by the server to authenticate the server. If the server cannot be authenticated, the user is notified of the problem and informed that an encrypted and authenticated connection cannot be established. If the server can be successfully authenticated, the client proceeds to the next step.

Using the data generated during the handshake so far, the client, which can use the cooperation of the server, depending on the cipher that is used, creates the pre-master secret for the session, encrypts it with the server's public key, obtained from the server's certificate, and sends the encrypted pre-master secret to the server.

If the server has requested authentication from the client, which is an optional step in the handshake, the client also signs another piece of data that is unique to this handshake and known by both the client and server. In this case, the client sends both the signed data and the client's own certificate to the server along with the encrypted pre-master secret.

If the server has requested client authentication, it tries to authenticate the client. If the client cannot be authenticated, the session is terminated. If the client can be successfully authenticated, the server uses its private key to decrypt the pre-master secret and then performs a series of steps, which is also performed by the client, to generate the master secret.

Both the client and the server use the master key to generate session keys, which are symmetric keys used to encrypt and decrypt information exchanges during the SSL session, and to guarantee their integrity, that is, to detect any changes in data from the moment they were sent until when they were received over the SSL connection.

The client sends a message to the server informing it that future messages from the client will be encrypted with the session key. It then sends a separate encrypted message, indicating that the client part of the handshake has ended.

The server sends a message to the client stating that future messages from the server will be encrypted with the session key. It then sends a separate encrypted message, indicating that the server part of the handshake has come to an end. The SSL handshake is now complete, and the SSL session has started. The client and server use session keys to encrypt and decrypt the data they send to each other and to validate their integrity.

Both client and server authentication involve encrypting some piece of data with one of the keys, of a public–private key pair, and decrypting it with another.

In the case of server authentication, the client encrypts the pre-key with the server's public key. Only the corresponding private key can properly decrypt the key; so the client has some confidence that the identity associated with the public key is, in fact, the server to which the client is connected. Otherwise, the server cannot decrypt the pre-key and cannot generate the necessary symmetric keys for the session, and the session will be terminated.

In the case of client authentication, the client encrypts random data with its private key, that is, it creates a digital signature. The public key in the client's certificate can correctly validate the digital signature only if the corresponding private key has been used. Otherwise, the server is unable to validate the digital signature and the session is ended.

9.4.1 Server Authentication

As explained for SSL handshake, the server sends the certificate to the client to authenticate itself. The client uses the certificate to authenticate the identity that the certificate claims to represent. To authenticate the connection between a public key and the server identified by the certificate containing the public key, an SSL-enabled client must receive a positive answer to the four questions shown in Figure 9.2.

A client that is SSL-enabled goes through the following steps to authenticate the server's identity:

1. Is the current date within the validity period? The client checks the validity period of the server certificate. If the current date and time are outside the period, the authentication process does not proceed. If they are within the validity period of the certificate, the customer proceeds to the next step.

2. Is the certificate authority (CA) issued valid? Each SSL-enabled client maintains a list of trusted CA certificates. This list determines which server certificates the client will accept. If the issuing CA's distinguished name (DN) matches the ND of a CA on the client's list of trusted CAs, the answer to that question is yes, and the client proceeds to the third stage. If the issuing CA is not on the list, the server will not be authenticated unless the client can verify a certificate chain ending in a CA that is on the list.

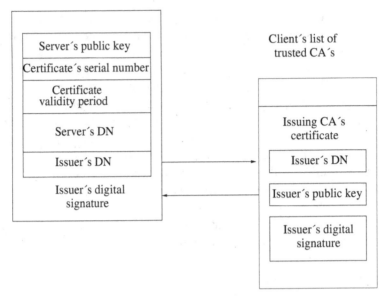

Figure 9.2   Server authentication procedure.

3. Does issuing the CA's public key validate the issuer's signature? The client uses the CA certificate's public key to validate the CA's digital signature on the server certificate being presented. If the information on the server's certificate has been modified since it was signed by the CA or if the CA's certificate key does not match the private key used by the CA to sign the server's certificate, the client will not authenticate the server's identity. At this point, the client has determined that the server's certificate is valid.

4. Is the domain name specified in the server's ND consistent with the actual domain name of the server? Although this question is technically not part of the SSL protocol, this requirement is the client's responsibility, and it gives some security as to the identity of the server, being the only way to protect against an attack known as "man in the middle." This step confirms that the server is actually located at the same address specified by the domain name in the server's certificate. Clients must perform this step and must refuse server authentication and not establish a connection if the domain name does not match. If the server's actual domain matches the domain name on the server's certificate, the client proceeds to step 5.

5. The server is authenticated, and the client continues with SSL handshake. If the client does not reach the fifth stage for any reason, the server identified by the certificate cannot be authenticated, and the user will be notified of the problem and informed that an encrypted and authenticated connection cannot be established. If the server requires client authentication, it performs the steps described in client authentication.

After all measures are taken, the server must use its private key to decrypt the pre-key that the client sends in the fourth step of SSL handshake. Otherwise, the SSL session will be terminated. This provides additional assurance that the identity associated with the public key in the server's certificate is, in fact, the server to which the client is connected.

9.4.2 Attack of the Man in the Middle

In this case, the encrypted information exchanged at the beginning of the SSL handshake is actually encrypted by the public or private key of the malicious program, rather than the real keys of the server or client. The malicious program ends up establishing a set of session keys for use with the real server and a different set of keys for use with the client.

This allows the malicious program to not only read all the data that is exchanged but also to change the data without being detected.

Therefore, it is important for the client to verify that the domain name in the server certificate matches the domain name of the server with which it is trying to communicate, in addition to verifying the validity of the certificate by performing the other steps described in server authentication.

9.4.3 Client Authentication

Servers enabled to use SSL can be configured to require client authentication, or cryptographic validation by the server of the client's identity. When a server configured this way requests client authentication, the client sends the server both a certificate and a separate piece of digitally signed data to authenticate itself. The server uses the digitally signed data to validate the public key in the certificate and to authenticate the identity that the certificate claims to represent.

The SSL protocol requires the client to create a digital signature by creating a one-way hash from data generated randomly during the handshake and known only to the client and server. The hash of the data is then encrypted with the private key that corresponds to the public key in the certificate being presented to the server.

To authenticate the connection between the public key and the person or entity identified by the certificate containing the public key, an SSL-enabled server must receive a positive answer to the previous four questions. Although the fifth question is not part of the SSL protocol, SSl servers can be configured to support this type of request in order to take advantage of user input in an LDAP directory as part of the authentication process.

An SSL-enabled server goes through the following steps to authenticate a user's identity:

1. Does the user's public key validate the user's digital signature? The server checks that the user's digital signature can be validated with the public key in the certificate. If the answer is positive, the server has established that the public key asserted to belong to someone matches the private key used to create the signature and that the data has not been tampered with since it was signed.

   At this point, however, the connection between the public key and the DN, specified in the certificate, has not yet been established. The certificate might have been created by someone attempting to impersonate the user. To validate the connection between the public key and the DN, the server must also complete the third and fourth steps.

2. Is the current date within the validity period? The client checks the validity period of the server certificate. If the current date and time are out of date, the authentication process does not proceed. If they are within the validity period of the certificate, the server moves to the third step.

3. Is the issuing CA a trusted CA? Each SSL-enabled server maintains a list of trusted CA certificates. This list determines which certificates the server will accept. If the DN of the issuing CA matches the DN of a CA on the server's list of trusted CAs, the answer to this question is yes, and the server proceeds to the fourth step. If the issuing CA is not on the list, the client will not be authenticated unless the server can verify a certificate chain ending in a CA that is on the list. Administrators can control which certificates are trusted or not trusted within their organizations by controlling the lists of CA certificates maintained by clients and servers.

4. Does issuing the CA's public key validate the issuer's signature? The server uses the CA certificate's public key to validate the CA's digital

signature on the client's certificate being presented. If the information in the client's certificate has been modified since it was signed by the CA or if the CA's certificate key does not match the private key used by the CA to sign the client's certificate, the server will not authenticate the client's identity.

If the CA's digital signature can be validated, the server treats the user's certificate as a valid "letter of introduction" from that CA and proceeds. At this point, the SSL protocol allows the server to consider the client authenticated and proceed with the connection as described in the sixth step. Netscape servers may optionally be configured to take the next two steps.

1. Is the user's certificate listed in the LDAP entry for the user? This optional step provides one way for a system administrator to revoke a user's certificate even if it passes the tests in all the other steps. The Netscape certificate server can automatically remove a revoked certificate from the user's entry in the LDAP directory. All servers that are set up to perform this step will then refuse to authenticate that certificate or establish a connection. If the user's certificate in the directory is identical to the user's certificate presented in the SSL handshake, the server goes on to step 6.

2. Is the authenticated client authorized to access the requested resources? The server checks what resources the client is permitted to access according to the server's access control lists (ACLs) and establishes a connection with appropriate access. If the server does not get to step 6 for any reason, the user identified by the certificate cannot be authenticated, and the user is not allowed to access any server resources that require authentication.

9.5 Data Protection with IPsec

The IPsec framework has three main components: authentication header (AH), encapsulating security payload (ESP), and Internet key exchange (IKE) IPsec adds integrity checking, authentication, encryption, and replay protection to IP packets. It is used for end-to-end security and also for creating secure tunnels between gateways. IPsec was designed for interoperability. When correctly implemented, it does not affect networks and hosts that do not support it and is independent of the current cryptographic algorithms and can accommodate new ones as they become available.

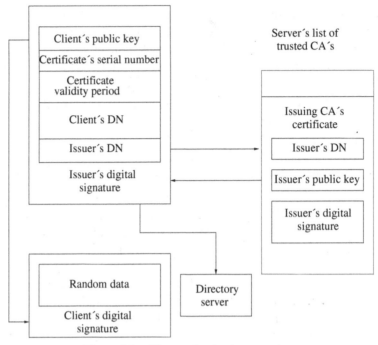

Figure 9.3   Client authentication procedure.

It works both with IPv4 and IPv6. In fact, IPsec is a mandatory component of IPv6. IPsec uses state-of-the-art cryptographic algorithms. The specific implementation of an algorithm for use by an IPsec protocol is often called a transform. For example, the DES algorithm used by ESP is called the ESP DES-CBC transform. The transforms, like the protocols, are published in the RFCs. Two major IPsec concepts should be clarified: security associations and tunneling.

9.5.1 Security Associations

A security association (SA) is a unidirectional (simplex) logical connection between two IPsec systems, uniquely identified by the following triple:
< Security Parameter Index, Destination IP Address, Security Protocol >

The definition of the members is as follows:

- Security parameter index (SPI) – A 32-bit value used to identify different SAs with the same destination address and security protocol. The SPI is carried in the header of the security protocol (AH or ESP). The SPI has only local significance, as defined by the creator of the SA. SPI values in the range 1–255 are reserved by the Internet Assigned Numbers Authority (IANA). Generally, the SPI is selected by the destination system during SA establishment.

- IP destination address – Can be a unicast, broadcast, or multicast IP address. However, currently SA management mechanisms are defined only for unicast addresses.
- Security protocol – Can be either an authentication header (AH) or an encapsulating security payload (ESP).

A security association can be in two modes, transport or tunnel, depending on the mode of the protocol for that SA. SAs are simplex; hence, for bidirectional communication between two IPsec systems, there must be two SAs defined, one in each direction.

A single SA gives security services to the traffic carried by it either by using AH or ESP, but not both. In other words, for a connection that should be protected by both AH and ESP, two SAs must be defined for each direction.

In this case, the set of SAs that defines the connection is referred to as an SA bundle. The SAs in the bundle do not have to terminate at the same endpoint. For example, a mobile host could use an AH SA between itself and a firewall and a nested ESP SA that extends to a host behind the firewall.

An IPsec implementation maintains two AS-related databases:

- Security policy database (SPD) specifies what security services are offered to the IP traffic, depending on factors, such as, source, destination, and whether it is inbound or outbound. It contains an ordered list of policy entries, separate for inbound and outbound traffic. These entries might specify that some traffic must bypass the IPsec processing, some must be discarded, and the remaining must be processed by the IPsec module. Entries in this database are similar to firewall rules or packet filters.
- Security association database (SAD) contains parameter information about each SA, such as AH or ESP algorithms and keys, sequence numbers, protocol mode and SA lifetime. For outbound processing, an SPD entry points to an entry in the SAD. That is, the SPD determines which SA is used for a given packet. For inbound processing, the SAD is consulted to determine how the packet must be processed. The user interface of an IPsec implementation usually hides or presents these databases in a friendly way.

9.5.2 Tunneling

Tunneling, or encapsulation, is a usual technique in packet-switched networks. It consists of wrapping a packet into a new one. That is, a new header is attached to the original packet. The entire original packet becomes the payload of another packet, as shown in Figure 9.4.

In general, tunneling is used to carry traffic of one protocol over a network that does not support that protocol directly. For example, NetBIOS or IPX can be encapsulated into IP to carry it over to TCP/IP WAN link (Tanenbaum, 2003).

In the case of IPsec, IP is tunneled by IP for a slightly different purpose: to allow full protection, including the encapsulated packet header. If the encapsulated packet is encrypted, an attacker cannot determine, for example, the destination address of that packet. The internal structure of a private network can be hidden in this way.

Tunneling requires intermediate processing of the original packet while en-route. The destination specified in the outer header, usually an IPsec firewall or router, receives the tunneled packet, extracts the original packet, and sends it to the ultimate destination. The processing overhead is compensated by the extra security.

A notable advantage of IP tunneling is the possibility to exchange packets with private IP addresses between two intranets over the public Internet, which requires globally unique addresses. Since the encapsulated header is not processed by the Internet routers, only the endpoints of the tunnel, the gateways, need to have globally assigned addresses. The hosts in the intranets can be assigned private addresses (for example, 10.x.x.x).

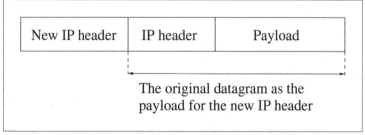

Figure 9.4   The original packet is the load of a new IP header.

9.5.3 Authentication Header

The authentication header (AH) is used to provide integrity and authentication to IP datagrams. Replay protection is also possible. Although its usage is optional, the replay protection service must be implemented by any IPsec-compliant system.

The services are connectionless, that is, they work on a per-packet basis. AH is used in two modes: transport mode and tunnel mode. AH authenticates as much of the IP datagram as possible. In transport mode, some fields in the IP header change en route and their value cannot be predicted by the receiver. These fields are called mutable and are not protected by AH.

The varying IPv4 fields are: type of service (TOS), flags, fragment offset, time to live (TTL), header checksum. When protection of these fields is required, tunneling should be used.

The payload of the IP packet is considered immutable and is always protected by AH. AH is identified by protocol number 51, assigned by the IANA. AH processing is applied only to non-fragmented IP packets. However, an IP packet with AH applied can be fragmented by intermediate routers. In this case, the destination first reassembles the packet and then applies AH processing to it.

If an IP packet that appears to be a fragment (offset field is non-zero, or the More Fragments bit is set) is input to AH processing, it is discarded. This prevents the so-called overlapping fragment attack, which misuses the fragment reassembly algorithm in order to create forged packets and force them through a firewall.

Packets that fail authentication are discarded and are not delivered to upper layers. This mode of operation reduces the chances of success for denial of service attacks, whose objective is to block the communication of a host or gateway by flooding it with packets.

9.5.4 Authentication Header Format

The fields of the authentication header format are illustrated in Figure 9.5 and described in the following.

The next header is an 8-bit field that identifies the type of what follows. The value of this field is chosen from the set of IP protocol numbers defined in the most recent assigned numbers RFC from the IANA. In other words, the IP header protocol field is set to 51, and the value that would have gone in the protocol field goes in the AH next header field.

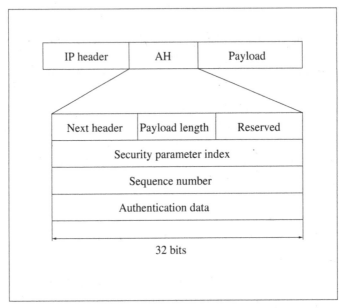

Figure 9.5   Authentication header (AH) fields.

The payload length is 8 bits long and contains the length of the AH header expressed in 32-bit words, minus 2. It does not report to the current payload length of the IP packet as a whole. If default options are used, the value is 4, which is equivalent to three 32-bit fixed words plus three 32-bit words of authentication data minus two. The reserved field is saved for future use. Its length is 16 bits and it is set to zero, as of now.

The security parameter index (SPI) field is 32 bits in length. It is used to identify different SAs with the same destination address and security protocol. The SPI is carried in the header of the security protocol (AH or ESP). Usually, the SPI is selected by the destination system during SA establishment.

The sequence number is a 32-bit field that acts as a monotonically increasing counter and is used for replay protection. Replay protection is optional but is a mandatory field. The sender always includes this field and it is at the discretion of the receiver to process it or not. At the establishment of an SA, the sequence number is initialized to zero. The first packet transmitted using the SA has a sequence number of 1. Sequence numbers are not allowed to repeat. Thus, the maximum number of IP packets that can be transmitted on any given SA is $2^{32}-1$. After the highest sequence number is used, a new

SA and, consequently, a new key is established. Anti-replay is enabled at the sender by default. If upon SA establishment the receiver chooses not to use it, the sender need not be concerned with the value in this field anymore. Typically, the anti-replay mechanism is not used with manual key management.

The authentication data is a variable length field containing the integrity check value (ICV) and is populated for 32 bits for IPv4 and 64 bits for IPv6. The ICV for each package is calculated with the algorithm selected at the start of the AS. As its name implies, it is used by the receiver to verify the integrity of the package on arrival. In theory, any MAC algorithm can be used to calculate the ICV.

The specification requires that HMAC-MD5-96 and HMAC-SHA-1-96 must be supported. The old RFDC 1826 requires Keyed MD5. In practice, Keyed SHA-1 is also used. Figure 9.6 shows the Keyed MD5 and Keyed SHA-1.

Implementations usually support two to four algorithms. When doing the ICV calculation, the mutable fields are considered to be filled with zero. Figure 9.7 shows the HMAC-MD5-96 and HMAC-SHA1-96.

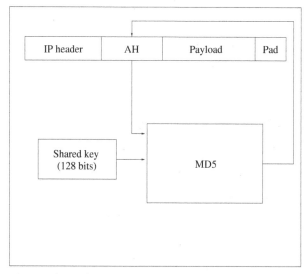

Figure 9.6   Keyed MD5 and Keyed SHA-1.

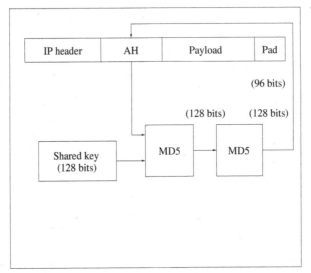

Figure 9.7   HMAC-MD5-96 and HMAC-SHA1-96.

9.5.5 Authentication Header in Transport and Tunnel Modes

In the transport mode, the authentication header is inserted immediately after the IP header. If the datagram already has IPsec header (s), then the AH is inserted before them. Transport mode is used by hosts and not by gateways. Gateways are not required to support transport mode. The advantage of transport mode is less processing overhead. The disadvantage is that varying fields are not authenticated. Figure 9.8 shows the authentication header in transport mode.

For the tunnel mode, the tunneling concept is applied, a new IP datagram is constructed, and the original IP datagram is converted into its payload. AH in transport mode is applied to the resulting datagram. Tunnel mode is used whenever either end of a security association is a gateway. Thus, between two firewalls, tunnel mode is always used. Gateways often also support transport mode. Figure 9.9 shows the authentication header in tunnel mode.

9.5.6 AH in Tunnel Mode

This mode is allowed when the gateway acts as a host, that is, in cases when traffic is destined to the gateway itself. For example, SNMP commands

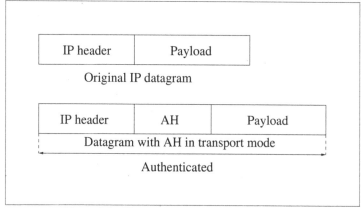

Figure 9.8 Transport mode authentication header.

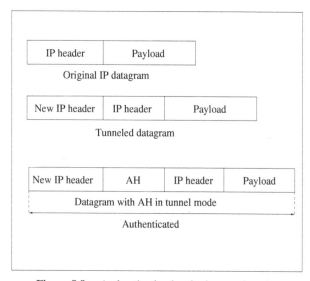

Figure 9.9 Authentication header in tunnel mode.

could be sent to the gateway using transport mode. In tunnel mode, the outer headers IP addresses are not necessarily the same as the inner headers addresses.

For example, two security gateways can operate an AH tunnel that is used to authenticate all traffic between the networks they connect together. This is a very typical mode of operation. The advantages of tunnel mode include total protection of the encapsulated IP datagram and the possibility of using private addresses. However, there is extra processing overhead associated with this mode.

9.6 Encapsulated Security Load

The encapsulating security payload (ESP) is used to provide integrity check, authentication, and encryption to IP datagrams. However, some restrictions apply:

- Integrity check and authentication are used together.
- Replay protection is selectable only in conjunction with integrity check and authentication.
- Replay protection can be selected only by the receiver.

Encryption can be selected independently of other services. It is highly recommended that, if encryption is enabled, integrity check and authentication be turned on. If only encryption is used, intruders could forge packets in order to mount cryptanalytic attacks. Although both authentication, with integrity check, and encryption are optional, at least one of them is always selected; otherwise, you would not be using ESP.

ESP is identified by protocol number 50, as assigned by the IANA. If both encryption and authentication with integrity check are selected, then the receiver first authenticates the packet and, only if this step is successful, proceeds with decryption. This mode of operation saves computing resources and reduces the vulnerability to denial of service attacks.

9.6.1 ESP Package Format

As shown in Figure 9.10, the payload data field is mandatory. It consists of a variable number of bytes of data described by the next header field. This field is encrypted with the cryptographic algorithm selected during SA establishment. If the algorithm requires initialization vectors, these are also included here. The ESP specification requires support for the DES algorithm in CBC mode (DES-CBC transform). Often, other encryption algorithms are also supported, such as triple-DES and CDMF, in the case of IBM products.

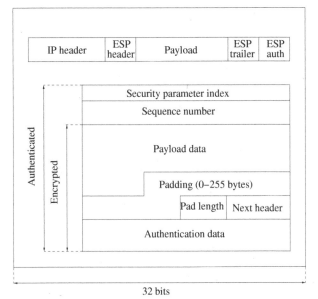

Figure 9.10   ESP package format.

Most encryption algorithms require that the input data must be an integral number of blocks. Also, the resulting ciphertext, including the padding, pad length, and next header fields, must terminate on a 4-byte boundary, so the next header field is right-aligned. For this reason, padding is included. It can also be used to hide the length, of the original messages. However, this could adversely impact the effective bandwidth. Padding is an optional field but necessary for some algorithms.

The pad length is an 8-bit field that contains the number of the preceding padding bytes. It is always present, and the value of 0 indicates no padding.

The next header is an 8-bit mandatory field that shows the data type carried in the payload, for example, an upper-level protocol identifier such as TCP. The values are chosen from the set of IP protocol numbers defined by the IANA authentication data This field is variable in length and contains the ICV calculated for the ESP packet from the SPI to the next header field inclusive.

The authentication data field is optional. It is included only when integrity check and authentication have been selected at SA initialization time. The ESP specifications require two authentication algorithms to be supported: HMAC with MD5 and HMAC with SHA-1. Often, the simpler keyed versions are also supported by IPsec implementations.

9.6.2 ESP in Transport Mode

In this mode, the ESP header is inserted right after the IP header. If the datagram already has IPsec header (s), then the ESP header is inserted before any of those. The ESP trailer and the optional authentication data are appended to the payload. ESP in transport mode provides neither authentication nor encryption for the IP header. This is a disadvantage since false packets might be delivered for ESP processing. The advantage of transport mode is the lower processing overhead. As in the case of AH, ESP in transport mode is used by hosts, not gateways. Gateways are not required to support transport mode.

9.6.3 ESP in Tunnel Mode

A new IP packet is constructed with a new IP header and ESP is then applied, as in transport mode. Since the original datagram becomes the payload data for the new ESP packet, it is completely protected, if both encryption and authentication are selected. However, the new IP header is still not protected. The tunnel mode is used whenever either end of a security association is a gateway. Thus, between two firewalls, the tunnel mode is always used. For example, two security gateways may operate an ESP tunnel which is used to secure all traffic between the networks they connect together.

Gateways often also support transport mode. This mode is allowed when the gateway acts as a host, that is, in cases when traffic is destined to the gateway itself. For example, SNMP commands could be sent to the gateway using transport mode. In tunnel mode, the outer header's IP addresses do not need to be the same as the inner headers' addresses. Hosts are not required to support tunnel mode. The advantages of the tunnel mode are total protection of the encapsulated IP datagram and the possibility of using private addresses. However, there is an extra processing overhead associated with this mode.

9.7 Espionage Between Countries

There is a mistake in thinking that the United States has only started to spy on Brazil in recent times. The book "Presence of the United States in Brazil," by Luis Alberto Moniz Bandeira, reports at least four attempts of American authorities to intervene or invade the country, throughout its history. Evidently, espionage preceded each of these attempts (Bandeira, 2007).

Espionage occurs even among preferred partners, like the United States and England. After Alan Turing helped breaking the code of the Enigma machine, invented by the Germans, and designing the Colossus computer to speed up decoding of messages during World War II, the English government hid the existence of the computer from the world and started using it to spy on Canada, Australia, and the United States, among others (Alencar, 2014a).

Monitoring of communications by American espionage and information services, mainly by the National Security Agency (NSA) and Central Intelligence Agency (CIA), always occurred but increased before the installation of the military regime in Brazil, after João Belchior Marques Goulart took office as president of the Country. Goulart was seen as a leftist, despite being a wealthy farmer with no ties to international socialism.

10

Theoretical Cryptography

"Cryptography succeeds when it's no longer the weakest link."

Ron Rivest, alluding to a cryptography maxim

10.1 Introduction

Cryptography is the basis for network security. It comes from the Greek word *kryptos*, for tomb, hidden or secret, combined with *graphein*, or writing. A free translation is hidden writing, which defines the proposal of ciphering a message. Cryptology combines the same previous prefix with *logia*, meaning study, to give the translation of study of secret coding.

Cryptography is the practice and study of techniques for secure communication in the presence of third parties. It is also a collection of techniques for construction and analysis of protocols to overcome the influence of jammers and which are related to various aspects of information security, such as data confidentiality, data integrity, authentication, and non-repudiation.

Along the centuries, governments used cryptography to cipher messages, which are supposed to remain in secrecy, while the spies tried to decipher them. During the wars, cryptography becomes more important, considering the nature of the operations.

Bletchley Park was the secret information and counter-information center in Great Britain during the Second World War. It was so secret that it remained as a legend after the end of the war, while England continued to break the codes of other countries, either enemies and friends (Haykin, 1999).

The first computer, the Collossus, was built in Bletchley Park, developed by Alan Mathison Turing (1912–1954), one of the greatest computer geniuses of the world. Destroyed after the end of the war, the Collossus was rebuilt from the original drawings.

189

The German Enigma code was broken there, with the help of Turing, which benefited from information passed by a Hungarian mathematician and a German officer who provided information to French spies in exchange for money. The French did not know how to decipher the code and passed the information on to the British.

10.2 Cryptographic Aspects of Computer Networks

The Internet has revolutionized the ways in which companies do business, since the Internet Protocol (IP) is undeniably efficient, inexpensive, and flexible (Alencar, 2012b). This section presents the main cryptographic aspects of modern TCP/IP computer networks, which include digital signature technology based on asymmetrical cryptographic algorithms, data confidentiality by applying symmetrical cryptographic systems, and system public key infrastructure (PKI) (Tanenbaum, 2003).

The possible vulnerabilities of TCP/IP computer networks, and possible techniques to eliminate them, are considered. It seems that only a general and multi-layered security infrastructure could cope with possible attacks to the computer network systems.

10.2.1 Potential Vulnerabilities of Computer Networks

The Internet Protocol (IP) is efficient, inexpensive, and flexible. However, the existing methods used to route IP packets leave them vulnerable to security risks, such as spoofing, sniffing, and session hijacking, and provide no form of non-repudiation for contractual or monetary transactions.

Organizations need to secure communications between remote offices, business partners, customers, and traveling and telecommuting employees, besides securing the internal environment. The transmission of messages over the Internet or intranet poses a risk, given the lack of protection at the existing Internet backbone.

Control and management of security and access between the entities in a company's business environment is important. Without security, both public and private networks are susceptible to unauthorized monitoring and access. Internal attacks might be a result of minimal or non-existent intranet security.

Risks from outside the private network originate from connections to the Internet and extranets. Password-based user access controls alone do not protect data transmitted across a network. Without security measures and controls in place, the data might be subjected to attack. Some attacks are passive, and the information is only monitored. Other attacks are active, and the information is altered to corrupt or destroy the data or the network itself.

10.3 Principles of Cryptography

The general model for an encryption system is shown in Figure 10.1. The original message, also called plaintext and denoted by the letter X, is converted to an apparently random sequence, called ciphertext and denoted by the letter Y. The encryption process consists of an encipherment algorithm that uses a key K, which is independent of the plaintext. The algorithm produces a different output depending on the key used (Stallings, 1999).

It is important to consider that a cryptanalyst, or hacker, could observe the encrypted message and try to decode it. The algorithm must, therefore, be powerful enough to resist the attack.

Encryption can be seen as a transformation, parameterized by K,

$$Y = E(X,K), \tag{10.1}$$

which converts the original text X to the ciphertext Y, to be transmitted.

At the reception side, the inverse transformation is performed on the ciphertext to produce the plaintext

$$X = D(Y,K). \tag{10.2}$$

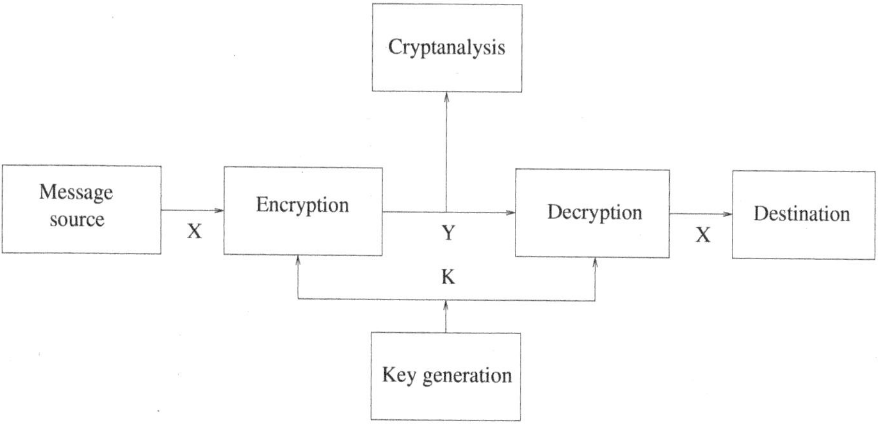

Figure 10.1 General model for a cryptosystem.

Of course, the destination must possess the key to be able to invert the transformation. Therefore, the transmission of the key is an important part of the cryptobusiness.

The security of a cryptosystem depends on certain assumptions:

- The encryption algorithm must be robust enough that it is difficult to decipher a message based only on the ciphertext.
- The key must be kept secret, and a secure channel should be provided for transmission.
- There is no need to keep the algorithm secret.

The application of cryptographic techniques depends on the knowledge of the security of the employed system. A cryptographic system can be classified as:

1. How to transform the plaintext to the ciphertext, as the encryption algorithms are based on two principles (Stallings, 1999):

    (a) Substitution, in which the individual symbols of the original text (bit, byte, letter, and words) are mapped into other elements.
    (b) Transposition, in which the elements of the original text are rearranged.

2. How the original text is processed, which involves a block cipher, that processes a block of plaintext at a time, or a convolutional cipher, also called stream cipher, that processes the input symbols as they arrive at the encryption sub-system.

3. How the key is generated, which can lead to the production of a single, or symmetric, key to be used by both the transmitter and receiver, or use different keys for the sender and receiver, also called asymmetric or public key encryption.

10.4 Information Theoretical Aspects of Cryptography

It is possible to use the terminology of information theory to describe a secure encryption system. The measure of information, or entropy, of a plaintext, considered as a subset of symbols selected from the set of all possible messages, is given by (van der Lubbe, 1997)

$$H(X) = -\sum_{l=1}^{L} p(x_l) \log p(x_l), \tag{10.3}$$

in which $p(x_l), l = 1, 2, \ldots, L$ represent the uncertainty of occurrence of the possible plaintexts, considering that the texts are mutually independent.

The entropy of the set of keys is given by

$$H(K) = -\sum_{m=1}^{M} p(k_m) \log p(k_m), \tag{10.4}$$

in which $k_m, m = 1,2, \ldots M$ are the possible keys.

By the same token, it is possible to introduce a measure of information for the ciphertext

$$H(Y) = -\sum_{n=1}^{N} p(y_n) \log p(y_n), \tag{10.5}$$

in which $p(y_n), n = 1, 2, \ldots, N$ represent the uncertainty of occurrence of the possible ciphertexts, considering that the texts are mutually independent. Usually, regarding the bijection from the plaintext into the ciphertext set, one considers $N = L$.

The conditional entropy, or key equivocation, is a measure of the uncertainty or information with respect to the key, when the ciphertext available is defined as

$$H(K|Y) = \sum_{m=1}^{M} \sum_{n=1}^{N} p(k_m, y_n) \log p(k_m|y_m). \tag{10.6}$$

The uncertainty with respect to the plaintext, or conditional entropy when the ciphertext is available, also known as message equivocation, is defined as

$$H(X|Y) = \sum_{l=1}^{L} \sum_{n=1}^{N} p(x_l, y_n) \log p(x_l|y_m). \tag{10.7}$$

The conditional entropy, or uncertainty with respect to the key, for a given plaintext and corresponding ciphertext is defined as

$$H(K|X,Y) = \sum_{m=1}^{M} \sum_{l=1}^{L} \sum_{n=1}^{N} p(k_m, x_l, y_n) \log p(k_m|x_l, y_m), \tag{10.8}$$

which is also known as key appearance equivocation.

Finally, the conditional entropy, or uncertainty with respect to the plaintext, when the key and the ciphertext are known, is defined as

$$H(X|Y,K) = \sum_{l=1}^{L} \sum_{n=1}^{N} \sum_{m=1}^{M} p(x_m, y_l, k_n) \log p(x_m|y_l, k_m). \tag{10.9}$$

Since there is a bijection between the plaintext and ciphertext sets, it is always true that

$$H(X|Y,K) = 0. \tag{10.10}$$

As expected, when the ciphertext and the key are available, it is possible to retrieve the plaintext correctly because the uncertainty with respect to X is null. There is no loss of information whatsoever at the receiver, and the full original message is recovered.

10.4.1 Relations Between the Entropies

A cipher system user requires that the key appearance equivocation, which is represented as $H(K|X,Y)$, be as high as possible because if a cryptanalyst manages to obtain both the plaintext and the ciphertext, then the uncertainty with respect to the key must be as large as possible.

This can be seen by the following argument. From the previous results in information theory, the joint measure of information in the plaintext, ciphertext, and key is

$$H(X,Y,K) = H(X|Y,K) + H(Y,K) = H(K|X,Y) + H(X,Y). \tag{10.11}$$

Also, consider the following relations:

$$H(Y,K) = H(K|Y) + H(Y) \tag{10.12}$$

and

$$H(X,Y) = H(X|Y) + H(Y). \tag{10.13}$$

Combining the equations, one obtains

$$H(X|Y,K) + H(K|Y) = H(K|X,Y) + H(X|Y). \tag{10.14}$$

But, as discussed, $H(X|Y,K) = 0$, which results in

$$H(K|Y) = H(K|X,Y) + H(X|Y), \tag{10.15}$$

which leads to

$$H(K|X,Y) = H(K|Y) - H(X|Y). \tag{10.16}$$

Therefore, to obtain a large key appearance equivocation, the message equivocation $H(X|Y)$ must be small. However, a small message equivocation implies that the uncertainty with respect to the plaintext, when the ciphertext is known, is small. But this is what must be avoided by the cryptosystem.

The uncertainty with respect to the key must be large to decrease the uncertainty with respect to the plaintext, and an increase in the uncertainty with respect to the plaintext decreases the uncertainty with respect to the key (van der Lubbe, 1997).

10.5 Mutual Information for Cryptosystems

The information theoretical approach provides additional results. The mutual information of the plaintext and the ciphertext is defined as

$$I(X;Y) = H(X) - H(X|Y) = H(Y) - H(Y|X). \qquad (10.17)$$

As expected, the objective of the cryptodesigner is to minimize the mutual information $I(X;Y)$. If the ciphertext provides no information about the original message, then

$$H(X|Y) = H(X),$$

and the mutual information between the plaintext and the encoded message is zero, or $I(X;Y) = 0$. This is referred to as the absolutely secure cryptosystem.

It is possible to obtain a lower limit for the mutual information between the plaintext and the ciphertext. First, take into account eqn (10.16), and recall that $H(K|X,Y) \geq 0$, then it follows that

$$H(K|Y) \geq H(X|Y). \qquad (10.18)$$

The use of one of the properties of entropy gives

$$H(K) \geq H(K|Y), \qquad (10.19)$$

and therefore,

$$H(K) \geq H(X|Y). \qquad (10.20)$$

Substituting this last inequality into eqn (10.17), gives

$$I(X;Y) \geq H(X) - H(K). \qquad (10.21)$$

Inequality 10.21 implies that a decrease in the uncertainty of a set of keys improves, on average, the independence between the plaintext and the ciphertext.

It is implicit in the derivation that absolute security of a cryptosystem can only be achieved for

$$H(K) \geq H(X). \tag{10.22}$$

For uniformly distributed keys,

$$H(K) = -\sum_{m=1}^{M} p(k_m) \log p(k_m) = \log(M), \tag{10.23}$$

and for uniformly distributed plaintexts,

$$H(X) = -\sum_{l=1}^{L} p(y_l) \log p(y_l) = \log(L). \tag{10.24}$$

Therefore,

$$H(K) = \log(M) \geq H(X) = \log(L), \text{ or } M \geq L, \tag{10.25}$$

because of the monotonicity of the logarithm function.

The last condition implies that the key must be at least of the same length as the message to be transmitted. However, the inequality can be avoided with the introduction of security events. When a security event happens, the system can be considered absolutely secure. If the event does not occur, the cryptosystem may not be fully secure (maurer.1989; massey.1990)

Even when there is a small chance of a security breach, an absolutely secure cipher is still feasible, for which $H(K) < H(X)$. A security event could be the transmission of a certain key K, followed by the plaintext, provided that the hacker does not type the secret key at the beginning. Without typing K, the sequence is obscure to the intruder, and the system is secure. If the hacker happens to type K, the ciphertext can be deciphered, and the system has become unreliable.

Therefore, the introduction of the concept of security event gives a new way of regarding the system security based on information theorems. Information security can be assured, even if the length of the key is smaller than the original sequence, given that a security event is defined for the system, and provided that the event actually occurs (van der Lubbe, 1997).

11

The Hash Function

"Codebreaking is an art, and codemaking is a science."

Adi Shamir

This chapter reviews the characteristics of the main digest algorithms, and presents a new derivation of the leftover hash lemma, using the collision probability to derive an upper bound on the statistical distance between the key and seed joint probability, and the hash bit sequence distribution.

11.1 Introduction to Network Security

The Internet has changed the way in which people interact and companies do business because the Internet Protocol (IP) is efficient, inexpensive, and flexible. However, the existing methods used to route IP packets leave them vulnerable to several security risks, such as spoofing, sniffing, and session hijacking, and do not provide the means of non-repudiation when dealing with contractual or monetary transactions.

Organizations need to secure communications between remote offices, business partners, customers, and traveling and telecommuting employees, and also secure the internal environment against intruders.

The transmission of messages over the Internet, even by intranets, to different entities implies a risk, given the lack of protection provided by the existing Internet backbone.

The control and management of security and access between different entities in a company business environment is certainly important.

Public and private networks are susceptible to unauthorized monitoring and access, and some internal attacks might be a result of minimal or non-existent intranet security.

Most risks to the private network originate from connections to the Internet, and password-based user access controls alone do not protect data transmitted across a network.

Some attacks are passive, that is, the information is only monitored, and other attacks are active because the information is altered with the objective of corrupting or destroying the data or disrupting the network.

The network communications, as a rule, occur in a plaintext (unencrypted) format, which allows an attacker who has gained access to data paths in a network to monitor and interpret (read) the traffic. When an attacker is eavesdropping on communications, it is referred to as sniffing or snooping.

After an attacker has read data, the next logical step is often to modify it. This can be done without the knowledge of the sender or receiver.

In general, the most important security problem that administrators face in an enterprise is the ability of an eavesdropper to monitor the network. Without strong encryption services, which are based on cryptography, the information can be read by attackers, as it traverses the network.

11.2 Hash Functions

The hash of the data is used as a digital fingerprint to sign and verify the information content. To prevent a security problem, the hash function must satisfy some criteria:

- The hash function must be collision-resistant, that is, it must be impractical to find a collision. No two different digital documents can be mapped onto the same hash format.
- The hash function must be a one-way function, that is, it must be difficult to find a pre-image for the function, for a given bit string, from the value range.

However, because the existence of collisions cannot be avoided, the requirements are of a theoretical nature.

A hash function maps a data file of arbitrary size onto a discrete sequence. The idea was introduced in cryptography, in 1976, by Diffie and Hellman, who identified the need for a one-way hash function as a building block of a digital signature scheme, as an algorithm to protect the authenticity of information. But, it is clear that the tool is useful to solve other security problems in communication and computer networks (Preneel, 1994).

The values returned by a hash function are called hash values, hash codes, or digests. The hash function is usually used in combination with a hash table, a data structure used for rapid data lookup. The hash function permits to speed up a database lookup by detecting duplicated records in a large file.

In mathematics, engineering, computing, and cryptography, universal hashing refers to the process of selecting a hash function at random, from a family of hash functions with a certain mathematical property. This guarantees a low number of collisions in average, even if the data is chosen by an adversary.

A family of functions $H = \{h : U \to [m]\}$ is called a universal family if, $\forall x,y \in U, x \neq y : \; \Pr_{h \in H}[h(x) = h(y)] \leq \frac{1}{m}$.

The privacy amplification permits the extraction of secret information, probably to be used as a cryptographic key, from a large data volume, that is only partially secret. The privacy amplification allows a large set of applications, according to the key (Maurer, 1995).

The next sections present a new derivation of the leftover hash lemma based on the Rényi entropy of order 2, also known as collision probability, to derive an upper bound on the statistical distance between the key and seed joint probability, and the hash bit sequence distribution.

11.3 Objectives of the Hash Function

A cryptographic hash function is used to verify whether a data file maps onto a certain hash value. On the other hand, it is difficult to reconstruct the information based on the hash value. Therefore, it is used to assure data integrity and is the building block of a hash-based message authentication code (HMAC), which provides message authentication.

The ideal cryptographic hash function has some desired properties. It is a deterministic function, that is, identical messages result in the same hash. It is fast to compute the hash value for a given message. It is impracticable to generate a message from its hash value, except by trying all possible messages; therefore, the hash is a one-way function.

Additionally, a small change to a message should cause an avalanche effect, that is, it changes the digest so considerably that the new hash value seems uncorrelated with the old hash value. It is collision resistant, that is, it is impractical to find two different messages with the same hash value. A cryptographic hash function should resist attacks on its pre-image.

The hash function creates a unidirectional process, that makes it impossible to guess the original contents of a file based only on the message digest. In the following, there is a short description of some known commercial algorithms (Wikipedia contributors, 2019).

Message-digest algorithm 4 (MD4), an algorithm developed between 1990 and 1991 by Ron Rivest has suffered several attacks and has been considered an insecure algorithm. It is described in RFC 1320; MD5, described in RFC 1321.

The message-digest algorithm 5 (MD5) is a hash algorithm with 128 bits, developed by RSA Data Security, Inc. and described in RFC 1321. It is used in peer-to-peer (P2P) protocols for identity verification and logins. Attach methods have been published for the MD5. The algorithm produces a digest with 128 bits (16 bytes).

The secure hash algorithm 1 (SHA-1) algorithm was jointly developed by the National Institute of Standards and Technology (NIST) and by the National Security Agency (NSA) in the United States. It presented problems, and the new versions SHA-2 and SHA-3 are in use. The program produces a digest with 160 bits (20 bytes).

The secure hash algorithm 3 (SHA-3) was announced by NIST in 2015. It is a subset of the Keccak cryptographic family, developed by Guido Bertoni, Joan Daemen, Michael Peeters, and Gilles Van Assche. The algorithm produces digests with 224, 256, 384, and 512 bits. In the Bitcoin blockchain, the mining process is conducted by running SHA-256 hashing functions.

Whirlpool is a cryptographic hash function developed by Paulo S. L. M. Barreto and por Vincent Rijmen and adopted by the International Organization for Standardization (ISO) and by the International Electrotechnical Commission (IEC) as part of the ISO 10118-3 standard. The algorithm produces a digest with 512 bits (64 bytes)

The RACE Integrity Primitives Evaluation Message Digest (RIPEMD-160) represents a family of cryptographic hash functions developed in 1996 by Hans Dobbertin, Antoon Bosselaers, and Bart Preneel, from the research group COSIC, Katholieke Universiteit Leuven. The software produces a digest with 160 bits (20 bytes).

BLAKE2 was created in 2012 by Jean-Philippe Aumasson, Samuel Neves, Zooko Wilcox-O'Hearn, and Christian Winnerlein to replace algorithms MD5 and SHA-1.

11.4 Mathematical Preliminaries

Consider the random variable $X \in \mathcal{X}$ in a non-empty finite set and two probability distributions, P e Q, defined in this set. Define the statistical distance, a measure of distance between the referred probability distributions, as (Alencar and Assis, 2020)

$$\Delta[P,Q] = \frac{1}{2} \sum_{x \in \mathcal{X}} |P(x) - Q(x)| = \frac{1}{2} \sum_{l=1}^{L} |p(x_l) - q(x_l)|, \qquad (11.1)$$

in which $L = |\mathcal{X}|$ is the number of distinct symbols in the message or file and $p(x_l)$ and $q(x_l)$ represent the individual symbol probabilities, for the respective distributions P and Q.

The collision probability is defined as

$$P_C(X) = \sum_{x \in \mathcal{X}} p_X^2(x) = \sum_{l=1}^{L} p^2(x_l). \qquad (11.2)$$

The uniform distribution has a small coincidence index, or collision probability, $P_C(X) = 1/L$, and any other distribution has a larger index.

The Rényi entropy of order α ($\alpha > 0, \alpha \neq 1$) is defined as (Rényi, 1961)

$$H_\alpha(X) = \frac{1}{1-\alpha} \log \sum_{i=l}^{L} p^\alpha(x_l), \ \alpha \neq 1. \qquad (11.3)$$

For the case $\alpha = 2$, one obtains the Rényi entropy of order 2,

$$H_2(X) = -\log P_C(X) = -\log \sum_{i=l}^{L} p^2(x_l). \qquad (11.4)$$

The following development demonstrates an inequality that involves the collision probability and the statistical distance. From Formula (11.2), it is possible to obtain

$$\sum_{x \in \mathcal{X}} \left[p_X(x) - \frac{1}{L} \right]^2 = \sum_{1}^{L} \left[p_X^2(x) - 2p_X(x)\frac{1}{L} + \frac{1}{L^2} \right] \qquad (11.5)$$

$$= P_C(X) - \frac{1}{L}.$$

Considering that $Y \in \mathcal{Y}$ is a random variable defined as

$$Y = \left[p_X(x) - \frac{1}{L} \right], \tag{11.6}$$

its mean squared value is given by

$$E[Y^2] = \frac{1}{L} \left[P_C(X) - \frac{1}{L} \right] = \frac{LP_C(X) - 1}{L^2}. \tag{11.7}$$

Applying the inequality $E^2[Y] \leq E[Y^2]$, one obtains

$$E[Y] \leq \sqrt{E[Y^2]} = \frac{\sqrt{LP_C(X) - 1}}{L}. \tag{11.8}$$

From the definition of statistical distance, from Formula (11.1),

$$\Delta[P,U] = \frac{1}{2} \sum_{x \in \mathcal{X}} \left| p_X(x) - \frac{1}{L} \right| = \frac{LE[Y]}{2}, \tag{11.9}$$

in which $U \in \mathcal{U}$ represents the uniform distribution.

Substituting the result from (11.8), it is possible to obtain the main inequality

$$\Delta[P,U] \leq \frac{\sqrt{LP_C(X) - 1}}{2}, \tag{11.10}$$

which relates the statistical distance and the collision probability.

11.5 Privacy Amplification

The privacy amplification theorem is related to the production of universal hash functions, as described in the following (Berens, 2013). First, assume that a sequence of n bits has been generated, related to the random variable X, defined in the set $\mathcal{X}$ (Stinson, 1994a).

For non-empty finite sets, $\mathcal{X}$, $\mathcal{S}$ e $\mathcal{Z}$, consider that $X \in \mathcal{X}$ is a random variable that represents the transmitted message, that S is a random variable uniformly distributed in the set of seeds $\mathcal{S}$, and that $Z \in \mathcal{Z}$ is the resulting random variable $f : \mathcal{X} \times \mathcal{S} \mapsto \mathcal{Z}$. Then, f is called a universal hash function, if (Carter and Wegman, 1979)

$$P\left[f(x,S) = f(x',S) \right] \leq \frac{1}{|\mathcal{Z}|}, \tag{11.11}$$

in which $|\cdot|$ represents the cardinality, and the operation is valid for every choice of $x \neq x'$, which are elements of $\mathcal{X}$.

In other words, any two keys of the universe collide with probability at most $\frac{1}{|\mathcal{X}|}$ when the hash function f is drawn randomly from the set. This is exactly the expected probability of collision if the hash function assigned truly random hash codes to every key.

From a seed, represented by the random variable S, uniformity distributed in the ser $\mathcal{S}$, it is possible to generate a safe key $K = f(X,S)$, of length r, with the application of a universal hash function X and S as follows COMP: $f : \mathcal{X} \times \mathcal{S} \mapsto \{0,1\}^r$, in which $1 \leq r \leq \infty$.

Given that the hash function f and the random variable S are public, the question is to know if the produced key is, in fact, safe.

The following result, known as the leftover hash lemma establishes that, given that the entropy of the sequence of n bits of X is superior to the sequence of the r bits of the key, the key is supposed to have been uniformly generated (Stinson, 1994b)

$$\Delta \left[P_{K,S}, U_{\{0,1\}^r} \cdot P_S \right] \leq \frac{1}{2} \cdot \left[2^{-\frac{1}{2}[H_2(X)-r]} \right], \qquad (11.12)$$

in which $P_{K,S}$ is the joint distribution of the key and seed, P_S is the seed distribution $U_{\{0,1\}^r}$ is the hash sequence uniform distribution, of length r, and $H_2(X)$ is the Rényi entropy of order 2.

This result is also known as the privacy amplification theorem. In the following, there is a new derivation of the leftover hash lemma that uses the collision probability.

In order to obtain Inequality (11.12), consider the following inequality, obtained from the Rényi entropy, for the case $\alpha = 2$:

$$H_2(X) = -\log P_C(X) = -\log \sum_{i=l}^{L} p^2(x_l). \qquad (11.13)$$

This entropy is the negative of the logarithm of the coincidence index. It follows from the Jensen inequality (Alencar and Alencar, 2016) that the Rényi entropy of order 2, or collision entropy, is limited by the Shannon entropy

$$H_2(X) \leq H(X), \qquad (11.14)$$

with equality if and only if the probability distribution of X is uniform.

Substituting the adequate parameters, one obtains

$$\Delta\left[P_{K,S}, U_{\{0,1\}^r} \cdot P_S\right] \leq \frac{\sqrt{|\mathcal{K}|P_C(K,S) - 1}}{2}. \tag{11.15}$$

Recalling that the probability distribution associated with the key is uniform and that each key has length r, one has $|\mathcal{K}| = 2^r$. Also, substituting result (11.9) into (11.15), one obtains

$$\Delta\left[P_{K,S}, U_{\{0,1\}^r} \cdot P_S\right] \leq \frac{1}{2}\sqrt{2^r 2^{-H_2(K,S)} - 1}. \tag{11.16}$$

Considering that the mapping $f : \mathcal{X} \times \mathcal{S} \mapsto \{0,1\}^r$ cannot increase the entropy, then $H(K,S) \leq H(X)$, which, after substitution into (11.16) gives

$$\begin{aligned}
\Delta\left[P_{K,S}, U_{\{0,1\}^r} \cdot P_S\right] &\leq \frac{1}{2} \cdot \left[2^{r-H_2(K,S)} - 1\right]^{1/2} \tag{11.17}\\
&\leq \frac{1}{2} \cdot \left[2^{r-H_2(K,S)}\right]^{1/2},
\end{aligned}$$

recalling that the probability is always positive or null.

Rewriting the inequality, exchanging the positions of the Rényi entropy, and the number of symbols r, with the resulting change in the signal, one obtains the result,

$$\Delta\left[P_{K,S}, U_{\{0,1\}^r} \cdot P_S\right] \leq \frac{1}{2} \cdot \left[2^{-\frac{1}{2}[H_2(K,S)-r]}\right]. \tag{11.18}$$

11.6 Conclusion

This chapter discussed the use of the hash function in cryptography and presented a new derivation of the upper bound on the statistical distance between the joint distribution of the key and the seed and the distribution of the hash bit distribution based on the collision probability.

12

Criminal Cases

"The salient and undeniable truth about cryptography is that no measure of violence or proscriptive legislation will ever solve a math problem."

Jacob Riggs

12.1 Introduction to Cybercrimes

This chapter presents some interesting cases of cybercrimes that caused some commotion around the world. A cybercrime is a criminal activity that either targets or uses a computer, a computer network, or a networked device. In other words, a computer is the object of the crime or is used as an instrument to commit an offense.

12.2 Types of Cybercrimes

Usually, a cybercrime is committed by cybercriminals, also known as hackers, who want to make money. Most crimes are committed by novice hackers, but some cybercriminals use advanced techniques and are technically skilled. A cybercrime is also carried out by criminal organizations.

It is usual to find political or personal objectives in those crimes, but a cybercrime rarely targets to damage computers for reasons other than profit. In the following, there are some specific examples of the different types of cybercrime (Kaspersky Laboratory, 2021):

- identity fraud, in which personal information is stolen to be used by the criminals;
- e-mail and Internet fraud, including spam;
- appropriation of financial or credit card data;
- theft and sale of corporate data;

- cyberextortion, when the criminal demands money to prevent a threatened attack;
- ransomware attack, a type of cyberextortion in which the criminal demands money to release the control of a computer facility;
- cryptojacking, in which hackers mine cryptocurrency using resources from others;
- cyberespionage, in which hackers access government information or company data;
- grooming, in which individuals make sexual advances to minors;
- phishing, the use of fake e-mail messages to obtain personal information from Internet users.

The possibilities to commit crimes over the Internet are limitless, but most cybercrimes are classified in two main categories:

- criminal activity that targets a specific computer or computer system and often involves viruses and other types of malware;
- criminal activity that uses computers to commit other crimes.

Several countries signed the European Convention of Cybercrime (ECC). The convention established a wide net of information and classified various malicious computer-related crimes that are considered cybercrime:

- illegally intercepting or stealing data;
- interfering with systems in a way that compromises a network;
- infringing copyright;
- illegal gambling;
- selling illegal items online;
- soliciting, producing, or possessing child pornography;

12.3 Famous Cybercrimes

This section presents some of the most famous cybercrimes that have been committed in this century. These crimes belong to three major categories: individual, property, and government (Mediacenter, 2021).

- Property – The hacker, for instance, steals a person's bank details to gain access to funds, make purchases online, or run phishing scams to convince people to give away their information or money. They could also use a malicious software to gain access to a web page with confidential information.

- Individual – In this type of cybercrime, one individual distributes malicious or illegal information online. This can include cyberstalking, pornography distribution, and trafficking.
- Government – This is an uncommon cybercrime, but is the most serious offense. A crime against a government is considered cyber terrorism. Government cybercrime includes hacking government websites, military websites, or distributing propaganda. The perpetrators are usually terrorists or spies from other governments.

12.3.1 The Mythical Pegasus

Pegasus is a spyware developed by a Israeli cyberarms firm founded by Niv Carmi, Omri Lavie, and Shalev Hulio, also called the NSO Group. It can be covertly installed on mobile phones and other devices that run versions of iOS and Android. The 2021 Project Pegasus revelations suggest that the current Pegasus software can exploit all recent iOS versions up to iOS 14.6.

Pegasus, which became operational in 2016, is capable of reading text messages, tracking calls and location, collecting passwords, accessing the target device's microphone and camera, and harvesting information from several software applications. The spyware, a Trojan horse that can infect cellphones, is named after the mythical winged horse Pegasus (Wikipedia contributors, 2021).

The actual number of devices that had been targeted was not revealed, but forensic analysis of 37 of the cellphones indicated that some successful hacks occurred. The list included 10 prime ministers, three presidents, a king, and people close to Saudi journalist Jamal Khashoggi, who was murdered in 2018 while visiting the Saudi consulate in Istanbul, Turkey (Joe Tidy, 2021).

12.3.2 Hackers' Attack to Colonial and JBS

A classical ransomware was committed by cybercriminals who forced the Colonial Pipeline to shut down, leading to a gas shortage on the East Coast of the United States, and warnings to not hoard gas in plastic bags. Then, a separate criminal group disrupted the meat supply chain by hitting the world's biggest food processing company, the Brazilian enterprise JBS, which is the acronym for José Batista Sobrinho, its founder.

A ransomware is a piece of software, usually spread through infected websites or phishing links. Once infected, the files stored on a PC become encrypted. A digital ransom letter followed suit, demanding a payment in Bitcoin to unlock the files.

Against the Federal Bureau of Investigation (FBI) advice, both Colonial and JBS paid millions of dollars in bitcoin as ransom to the hacker, the Russia-based REvil Group. Intelligence officials are worried about the proliferation of ransomware attacks, which they now consider as much of a priority as terrorism (Wilde et al., 2021b).

12.3.3 The Biggest Ransomware Attack Ever

In 2021, a cybercrime organization infected one million systems across 17 countries and demanded US$ 70 million in bitcoin, in exchange for a universal decryptor that would return users' access.

The hackers targeted the United States information technology (IT) firm Kaseya and then used that company's software to invade the victims' systems, which they held hostage. They timed the attack for the Fourth of July weekend.

Most of the victims were public agencies and businesses like Swedish grocery chain Coop, which had to close most of its 800 stores during the weekend. This was the Russia-based REvil Group's second widespread attack of the year (Wilde et al., 2021a).

12.3.4 Credit Cards Data Stolen from Home Depot's System

In 2014, retailer Home Depot's system was breached, exposing data from over 50 million credit cards. The thieves used a vendor's username and password to get onto the company's computer network and then installed malware on its point-of-sale systems, which meant that consumers swiping their credit cards were literally handing over their data to the criminals (Initiative, 2017).

12.3.5 Largest Theft of Customer Data

Bank JPMorgan Chase, in 2014, disclosed a massive breach that compromised the data of 76 million households and seven million small businesses. Other United States financial institutions, brokerage firms, and financial news publishers were targeted, including Citigroup, HSBC, Dow Jones, and payroll service company ADP.

Three men, now under arrest and pending trial, had set up a scheme of hacking as a business model, according to Preet Bharara, the United States attorney for the Southern District of New York. He called the breach: "The single largest theft of customer data from a United States financial institution ever."

The men were charged with use of the stolen information in pump-and-dump schemes to manipulate stock prices by sending fake e-mails to customers whose data was stolen, tricking them into investing and then profiting by the rise in stock price. The three men also allegedly operated unlawful Internet gambling sites, distributed counterfeit and malicious software, and operated an illegal Bitcoin exchange.

12.3.6 Yahoo Data Breach

Yahoo! has been the target of at least two major breaches. In 2016, the company disclosed that it had a 2014 breach that compromised the data of at least 500 million users.

Later on, the company reported that another breach occurred, in 2013, that exposed data of more than a billion Yahoo! users. The company did not explain the reason for the delay in reporting the breaches, which could cause problems with regulators.

The United States Securities & Exchange Commission issued guidance, in 2011, that required companies to disclose material information about cyber incidents if they could impact investors. The agency is reportedly investigating the company.

12.3.7 Customer's Data Stolen from Healthcare Companies

The past years have shown that consumer health data can be vulnerable, as hackers increasingly target health insurance and medical information.

In 2015, three healthcare companies, Anthem, Premera Blue Cross, and CareFirst BlueCross BlueShield, were hacked. Anthem's was the largest, exposing data from 79 million customers. Premera had information on more than 11 million customers stolen by hackers. CareFirst uncovered a breach that had compromised the information of over a million customers.

12.3.8 Social Security Numbers Stolen from IRS

In 2015, the United States Internal Revenue Service (IRS) had a breach that exposed more than 700,000 social security numbers and other sensitive information. Published reports say that the hackers took advantage of an online IRS program called Get Transcript, which allowed taxpayers to access their tax history.

The thieves used stolen social security numbers to fraudulently file for refunds. According to a report by the inspector general, in the 2016 tax season, the IRS identified 42,148 tax returns with US$ 227 million claimed in fraudulent refunds, and that was only for the first quarter of that year.

12.3.9 Government Data Breach Due to Outdated Technology

The United States Office of Personnel Management (OPM) exposed records of as many as 21.5 million people, one of the largest breaches of government data in United States history.

The exposed information included social security numbers, names, dates and places of birth, health and financial details, and even fingerprints of people who had been subjected to government background checks.

A congressional report published in September 2016 said the government was using outdated technology, which left its systems vulnerable. One of the hackers used a contractor's credentials to log on, install malware, and create a backdoor to the government network.

12.3.10 Google Corporate Servers Hacked in China

In 2009, hackers accessed several of Google's corporate servers in China, stealing intellectual property and other information. The company said it had sufficient evidence to suggest that the main objective of the attackers was to access the Gmail accounts of Chinese human rights activists.

In 2013, United States government officials said the Chinese hackers had accessed a sensitive database that contained court orders authorizing surveillance, possibly of Chinese agents who had Gmail accounts.

A Microsoft official suggested that Chinese hackers had targeted its servers, at about the same time as Google's, possibly searching for similar information about its e-mail service.

12.3.11 Sony Pictures Hacked

In 2014, hackers attacked the computer network of Sony Pictures, stealing employee e-mails, information on executive salaries, and copies of unreleased movies. There was widespread speculation that the group was trying to disrupt release of the film, The Interview, a comedy depicting a plot to assassinate North Korean leader Kim Jong-un.

The United States government blamed the North Korean government for the breach, the first time the United States government publicly accused a country of a cyberattack.

12.3.12 Wikileaks Obtained Sensitive Information from the Democratic National Committee

In 2016, Wikileaks published a series of e-mails taken from servers of the Democratic National Committee (DNC). The e-mails contained private correspondence, some of which ridiculed the campaign of the Bernie Sanders, and sensitive financial data on high-profile donors to Hillary Clinton's campaign.

The revelations prompted the resignation of the DNC's chairperson and arguably impacted the United States election. The intelligence agencies said they were confident that the Russian government was behind the hacks and even issued a report at the end of 2016 providing details on how the Russians allegedly carried out the exploit.

12.3.13 A DDoS Attack that Took Down Twitter, PayPal, and Netflix

The year 2016 was the first time the Internet of Things (IoT) was widely used in a cybercrime. A cyberattack on one of the companies that host the Internet's Domain Name System (IDNS), a directory of Internet addresses, took down many of the Internet's most popular sites, including Twitter, Netflix, Paypal, and Spotify.

The attack was of a common type, called a distributed denial of service (DDoS), which aims at shutting down systems by bombarding them with several requests at the same time. The unusual aspect of the attack was that, rather than using zombie personal computers (PC), which are used to run malware, transforming them into robots that send the requests, the attackers used common things that are connected to the Internet, such as baby monitors and digital recorders.

The company, called Dyn, said the assault came from millions of Internet addresses, making it one of the largest cyberattacks of all time. Experts believe that, as more things are connected to the Internet, the cybercrime will get worse.

12.3.14 MafiaBoy's Teenage Crimes

Michael Calce, known online as MafiaBoy, was only 15 years old when he took the world by storm with a series of nasty DDoS attacks. Focusing mainly on large corporations, MafiaBoy managed to take down CNN, eBay,

Amazon, and Yahoo, which was the World's largest search engine at the time (Mangion.2019)

The attacks served to wake up the United States government, causing President Clinton to convene a cyber security task-force. The highest estimate for the damage caused by MafiaBoy comes to US$ 1.2 billion; however he only suffered a relatively minor punishment because he was a minor at the time, which included eight months of open custody, a small fine, and a year of probation.

12.3.15 Epsilon Attacked by Hackers

One of the more extensive data breaches of all time happened in 2011 when Epsilon, an e-mail marketing company, had its database of client e-mail addresses stolen by hackers. Epsilon had around 2200 corporate clients at the time and was ultimately responsible for sending out more than 40 billion e-mails per year.

All the individual e-mail addresses were compromised by the breach, and the chances of spear phishing attacks, a more focused version of phishing scams, increased. It is difficult to estimate the full extent of the damage caused by the Epsilon hack, but experts said the figure could have gone up to US$ 4 billion.

12.3.16 NHS Patients Had Their Computers Infected

In 2017, a particularly infectious form of ransomware called WannaCry was let loose upon computer networks worldwide, creating chaos everywhere. In a matter of days, it had infected almost 200,000 devices in 150 countries. Once infected, the files stored on a PC would become encrypted. A digital ransom letter would then appear on the desktop, demanding a payment in Bitcoin to unlock the files.

Apart from the many private users who were affected, WannaCry also caused damage to several large corporations, such as the National Health System (NHS), Renault, Nissan, and FedEx. In the case of the NHS, as many as 70,000 devices may have been infected, including MRI scanners, theater equipment, and blood-storage fridges, which resulted in over 19,000 appointment cancelations, immediately costing around US$ 30 million and an additional US$ 100 million in subsequent repairs and upgrades.

12.3.17 When the United States Department of Defense was Attacked

In 1998, the United States Department of Defense (DoD) suffered a series of attacks on its network via a well-known operating system vulnerability. After gaining access to each network, the hackers implanted data gathering devices (sniffers) to collect data to be retrieved later. Networks compromised included the US Air Force, US Navy, US Marines, NASA, and the Pentagon, granting the hackers access to hundreds of network passwords.

Once detected, the United States government initially suspected that hackers from Iraqi were behind the breach since the Unites States was preparing for possible military action against Iraq at the time. However, they eventually discovered that the attacks were conducted by three teenage hackers, two from California and one from Israel.

12.3.18 The Ashley Madison Hack

The Ashley Madison hack, in 2015, was not a large cybercrime in terms of financial damage caused or number of victims involved, but it caused many problems because of the sensitivity of the data stolen. Ashley Madison offers a dating service for cheating spouses, which is why the 300 GB of user data leaked not only included users' banking data and real names but also details of their private sexual desires.

The group of hackers, known as Impact Team, said the website's cyber security was almost non-existent. They also exposed the company for failing to delete users' data, after charging them to do so. In the end, the company got off rather lightly and only had to settle two dozen class-action lawsuits for US$ 11.2 million, plus a US$ 1.66 million fine for the data breach.

12.3.19 The Stuxnet Worm Attack

The Stuxnet worm was one of the world's first instances of weaponized computer code. This means that it was not only able to cause damage digitally but could also cause physical harm to targets in the real world.

Stuxnet was most famously used against a nuclear research facility in Tehran, in 2012. The worm exploited four zero-day flaws within the research facility's system, infected over 60,000 state-owned computers, and physically destroyed approximately 1000 nuclear centrifuges. This was around a fifth of the total owned by Iran and slowed nuclear projects by a number of years.

12.3.20 The Iceman Attack

Perhaps, one of the most ambitious individuals in hacking history, the Iceman, codename of Max Ray Butler, stole the details of more than two million credit cards over the course of his criminal career. What makes him particularly unique is the fact that his victims were not just businesses, banks, and individuals but also rival hackers themselves.

Operating via the Deep Web, Butler set up CardersMarket, a forum for cyber-criminals to buy, sell, and exchange illicitly obtained personal details, such as credit card information. However, this scheme was not enough for him, and he then hacked into another similar site before permanently merging it with his own.

12.3.21 Bitcoin Stolen from Bitfinex Exchange in Hong Kong

Nearly 120,000 units of digital currency bitcoin, worth about US$72 million, was stolen from the exchange platform Bitfinex in Hong Kong, rattling the global bitcoin community in the second-biggest security breach ever of such an exchange (Baldwin, 2016).

Bitfinex was, at the time, the world's largest dollar-based exchange for bitcoin, known in the digital currency community for having deep liquidity in the dollar and bitcoin currency pair. The volume of bitcoin stolen amounted to about 0.75% of all bitcoin in circulation, in 2016.

The security breach came two months after Bitfinex was ordered to pay a US$ 75,000 fine by the United States Commodity and Futures Trading Commission, in part for offering illegal off-exchange financed commodity transactions in bitcoin and other digital currencies.

12.3.22 Crypto Exchange Bithumb Hacked

According to a report from CoinDesk Korea, Bithumb may have lost US$ 6.2 million in the recent breach. The amount was moved from Bithumb's wallet on March 29, 2019. The exchange did not confirm or deny the report, at that time, in what was suspected to be an insider job (Zhao, 2019).

The possibility of Bithumb ever managing to retrieve the stolen funds may be slim, according to crypto security expert Cosine Yu, co-founder of the security firm SlowMist. The hacker has already managed to launder most of the stolen money, Yu said, meaning the assets have been transferred to a large number of addresses that are not necessarily owned by any exchanges.

The remaining assets in Bithumb's hot wallet were removed to its cold (offline) wallet, to prevent further losses until the manner of the breach could be identified and any vulnerabilities fixed.

12.3.23 Espionage Between Partners

Espionage occurs even among preferred partners, such as the United States and England. After Alan Turing helped breaking the code of the Enigma machine, invented by the Germans, and designing the Colossus computer to speed up decoding of messages during World War II, the British government hid the existence of the computer from the world, and started using it to spy on Canada, Australia, and the United States, among others. And the espionage continued until the 1960s (Alencar, 2014a).

Monitoring of communications by American espionage and information services, mainly by the National Security Agency (NSA) and Central Intelligence Agency (CIA), always occurred but increased before the installation of the military regime in Brazil, after João Belchior Marques Goulart took office as president of the country. Goulart was seen as a leftist, despite being a wealthy farmer with no ties to international socialism.

The United States has started to spy on Brazil a long time ago. The book "Presence of the United States in Brazil" by Luis Alberto Moniz Bandeira reports at least four attempts of American authorities to intervene or invade the country throughout its history. Evidently, espionage preceded each of these attempts (Bandeira, 2007).

Appendix A

Probability Theory

"Anyone who attempts to generate random numbers by deterministic means is, of course, living in a state of sin."

John von Neumann

A.1 Set Theory and Measure

Georg Cantor (1845–1918) developed the modern theory, of sets, in the end of the nineteenth century, established the mathematical and logical basis for the theory and demonstrated several important results. The concept of set cardinality, defined by Cantor, was fundamental to the theory. Cantor was born in Saint Petersburg but lived most of his academic life in the city of Halle, Germany (Boyer, 1974). The basic ideas of universal set, empty set, set partition, discrete systems, continuous systems, and infinity are as old as humanity itself.

Over the years, philosophers and mathematicians had tried to characterize the infinite, with no success. In 1872, J. W. R. Dedekind (1831–1916) indicated the universal property of infinite sets. He stated that a set is called infinite when it is similar to a part of itself; on the contrary, the set is finite (Boyer, 1974).

Cantor also investigated the properties of infinite sets, but, different from Dedekind, he noticed that the infinite sets are not the same. This led to the concept of cardinal numbers to establish a hierarchy of infinite sets in accordance with their respective powers. Cantor's ideas established the set theory as a complete subject. As a consequence of his published results on transfinite arithmetic, considered advanced for his time, Cantor suffered attacks from mathematicians like Leopold Kronecker (1823–1891), who barred him for a position at the University of Berlin.

Cantor found a position at the ancient and small University of Halle, in the medieval city of Halle in Germany, famous for its mines of rock salt, and died there in an institution for mental health treatment, following his attempts to apply his theory to justify religious paradigms in scientific events.

A.1.1 Basic Set Theory

The notion of a set, as simple as it can be, is axiomatic, in a sense that it does not admit a definition that does not resort to the original notion of a set. The mathematical concept of a set is fundamental for all known mathematics and is used to build important concepts, such as relation, Cartesian product, and function. It is also the basis for the modern measure theory, developed by Henry Lebesgue (1875–1941).

The set theory is based on a set of fundamental axioms: Axiom of Extension, Axiom of Specification, Peano's Axioms, Axiom of Choice, besides Zorn's Lemma and Schröder – Bernstein's Theorem (Halmos, 1960).

The objective of this section is to present the theory of sets in an informal manner, just quoting those fundamental axioms, since this theory is used as a basis to establish a probability measure. Some examples of common sets are given in the following.

- the binary set: $\mathbb{B} = \{0, 1\}$;
- the set of natural numbers, including zero: $\mathbb{N} = \{0, 1, 2, 3, \ldots\}$;
- the set of odd numbers: $\mathbb{O} = \{1, 3, 5, 7, 9, \ldots\}$;
- the set of integer numbers: $\mathbb{Z} = \{\ldots, -3, -2, -1, -2, 0, 1, 2, 3, \ldots\}$;
- the set of real numbers: $\mathbb{R} = (-\infty, \infty)$.

There are two really important relations in set theory: the belonging relation, denoted as $a \in A$, in which a is an element of the set A, and the inclusion relation, $A \subset B$, which is read "A is a subset of the set B," or B is a super set of the set A. Sets may also be specified by propositions. For example, the empty set can be written as $\emptyset = \{a \,|\, a \neq a\}$, i.e., the set the elements of which are different from themselves.

A universal set contains all other sets of interest. An example of a universal set is provided by the sample space in probability theory, usually denoted as S or Ω. The empty set is that set which contains no element, usually denoted as $\emptyset$ or $\{\ \}$. It is implicit that the empty set is contained in any set, i.e., that $\emptyset \subset A$, for any given set A. However, the empty set is not in general an element of any other set.

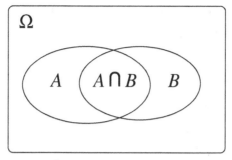

Figure A.1 A Venn diagram that represents two intersecting sets.

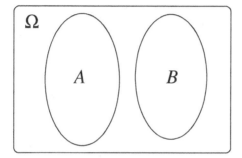

Figure A.2 A Venn diagram representing disjoint sets.

A usual way to represent sets is by means of the Venn diagram, as illustrated in Figure A.1.

Sets are said to be *disjoint* if they have no element in common, as illustrated in Figure A.2. Thus, for example, the set of even natural numbers and the set of odd natural numbers are disjoint.

A.1.2 Some Operations on Sets

It is possible to operate on sets to produce new sets or families of sets. The basic set operations are the complement, the union, the intersection, the subtraction, and the symmetric difference.

- The operation $\overline{A}$ represents the complement of A with respect to the sample space Ω.
- The union of two sets is composed of elements that belong to A or B and is written as $A \cup B$.
- The intersection of two sets is composed of elements that belong to A and B and is written as $A \cap B$.

- The subtraction of sets, denoted as $C = A - B$, gives, as a result, the set of elements that belong to A and do not belong to B.
  Note: If B is completely contained in $A : A - B = A \cap \overline{B}$.
- The symmetric difference is defined as the set of elements that belong to A and B but do not belong to $(A \cap B)$. It is written commonly as $A \triangle B = A \cup B - A \cap B$.

The generalization of these concepts to families of sets as, for example, $\cup_{i=1}^{N} A_i$ and $\cap_{i=1}^{N} A_i$ is immediate. The following properties are usually employed as axioms in developing the theory of sets (Lipschutz, 1968).

- **Idempotent**
  $$A \cup A = A, \qquad A \cap A = A$$
- **Associative**
  $$(A \cup B) \cup C = A \cup (B \cup C), \qquad (A \cap B) \cap C = A \cap (B \cap C)$$
- **Commutative**
  $$A \cup B = B \cup A, \qquad A \cap B = B \cap A$$
- **Distributive**
  $$A \cup (B \cap C) = (A \cup B) \cap (A \cup C),$$
  $$A \cap (B \cup C) = (A \cap B) \cup (A \cap C)$$
- **Identity**
  $$A \cup \emptyset = A, \qquad A \cap U = A$$
  $$A \cup U = U, \qquad A \cap \emptyset = \emptyset$$
- **Complementary**
  $$A \cup \overline{A} = U, \qquad A \cap \overline{A} = \emptyset \qquad \overline{(\overline{A})} = A$$
  $$\overline{U} = \emptyset, \qquad \overline{\emptyset} = U$$
- **de Morgan laws**
  $$\overline{A \cup B} = \overline{A} \cap \overline{B}, \qquad \overline{A \cap B} = \overline{A} \cup \overline{B}$$

A.1.3 Families of Sets

The concept of family is important to characterize finite or infinite combinations of sets. The increasing sequence of sets, such that $\lim_{i \to \infty} \cup A_i = A$, is one of the most useful families of sets. This sequence is used in proofs of limits over sets.

The decreasing sequence of sets is defined in a similar manner, as $\lim_{i \to \infty} \cap A_i = A$, and is also used in proofs of limits over sets.

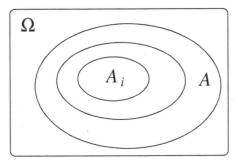

Figure A.3   Increasing sequence of sets.

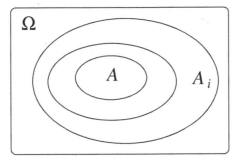

Figure A.4   Decreasing sequence of sets.

A.1.4 Indexing of Sets

The Cartesian product is useful to express the idea of indexing of sets. Indexing expands the possibilities for the use of sets and permits to generate new entities, such as vectors and signals.

Example: Consider $A_i = \{0,1\}$. Starting from this set, it is possible to construct an indexed sequence of sets by defining its indexing: $\{A_{i\epsilon I}\}$, $I = \{0,...,7\}$. This family of indexed sets A_i constitutes a finite discrete sequence, i.e., a vector.

Example: Again, let $A_i = \{0,1\}$, but now use $I = \mathbb{Z}$, the set of positive and negative integers plus zero. It follows that $\{A_{i\epsilon Z}\}$, which represents an infinite series of 0s and 1s, i.e., it represents a binary digital signal. For example, ...,0011111000 ....

Example: Using the same set $A_i = \{0,1\}$, but now indexing over the set of real numbers, $\{A_{i\epsilon I}\}$, in which $I = \mathbb{R}$, it is possible to form a signal which is discrete in amplitude but continuous in time.

Example: Considering $A = \mathbb{R}$ and $I = \mathbb{R}$, the resulting set represents an analog signal, that is, a signal that is continuous in time and in amplitude.

A.1.5 An Algebra of Sets

For the construction of an algebra of sets or, equivalently, for the construction of a field over which operations involving sets make sense, a few properties have to be obeyed.

1. If $A \in \mathcal{F}$, then $\overline{A} \in \mathcal{F}$. A is the set containing desired results, or over which one wants to operate.
2. If $A \in \mathcal{F}$ and $B \in \mathcal{F}$, then $A \cup B \in \mathcal{F}$.

The properties guarantee the closure of the algebra with respect to finite operations over sets. It is noticed that the universal set Ω always belongs to the algebra, i.e., $\Omega \in \mathcal{F}$, because $\Omega = A \cup \overline{A}$. The empty set also belongs to the algebra, i.e., $\emptyset \in \mathcal{F}$, since $\emptyset = \overline{\Omega}$ follows by Property 1.

Example: The family $\{\emptyset, \Omega\}$ complies with the above properties and, therefore, represents an algebra. In this case, $\emptyset = \{\}$ and $\overline{\emptyset} = \Omega$. The union is also represented, as can be easily checked.

Example: Given the sets $\{C_H\}$ and $\{C_T\}$, representing the faces of a coin, respectively, if $\{C_H\} \in \mathcal{F}$ then, $\{\overline{C_H}\} = \{C_T\} \in \mathcal{F}$. It follows that $\{C_H, C_T\} \in \mathcal{F} \Rightarrow \Omega \in \mathcal{F} \Rightarrow \emptyset \in \mathcal{F}$.

The previous example can be explained by the following argument. If there is a measure for heads, then there must be also a measure for tails, if the algebra is to be properly defined. Whenever a probability is assigned to an event, then a probability must also be assigned to the complementary event.

The cardinality of a finite set is defined as the number of elements belonging to this set. Sets with an infinite number of elements are said to have the same cardinality if they are equivalent, i.e., $A \sim B$ if $\sharp A = \sharp B$. Some examples of sets and their respective cardinals are presented next.

- $I = \{1, \cdots, k\} \Rightarrow C_I = k$;
- $\mathbb{N} = \{0, 1, \ldots\} \Rightarrow C_N$ or $\aleph_0$;
- $\mathbb{Z} = \{\ldots, -2, -1, 0, 1, 2, \ldots\} \Rightarrow C_Z$;
- $\mathbb{Q} = \{\ldots, -1/3, 0, 1/3, 1/2, \ldots\} \Rightarrow C_Q$;
- $\mathbb{R} = (-\infty, \infty) \Rightarrow C_R$ or $\aleph$.

For the above examples, the following relations are verified: $C_{\mathbb{R}} > C_{\mathbb{Q}} = C_{\mathbb{Z}} = C_{\mathbb{N}} > C_I$. The notation $\aleph_0$, for the cardinality of the set of natural numbers, was first employed by Cantor.

The cardinality of the power set, i.e., of the family of sets consisting of all subsets of a given set I, $\mathcal{F} = 2^I$, is 2^{C_I}.

A.1.6 The Borel Algebra

The Borel algebra, established by Félix Edouard Juston Émile Borel (1871–1956), and written as $\mathcal{B}$, or σ-algebra, is an extension of the algebra so far discussed to operate with limits at infinity. The following properties are required from a σ-algebra:

1. $A \in \mathcal{B} \Rightarrow \overline{A} \in \mathcal{B}$;
2. $A_i \in \mathcal{B} \Rightarrow \bigcup_{i=1}^{\infty} A_i \in \mathcal{B}$.

The properties guarantee the closure of the σ-algebra with respect to enumerable operations over sets. They allow the definition of limits in the Borel field.

Example: Considering the above properties, it can be verified that $A_1 \cap A_2 \cap A_3 \cdots \in \mathcal{B}$. In effect, it is sufficient to notice that

$$A \in \mathcal{B} \text{ and } \mathcal{B} \in \mathcal{B} \Rightarrow A \cup \mathcal{B} \in \mathcal{B},$$

and

$$\overline{A} \in \mathcal{B} \text{ and } \overline{\mathcal{B}} \in \mathcal{B} \Rightarrow \overline{A} \cup \overline{\mathcal{B}} \in \mathcal{B},$$

and, finally,

$$\overline{\overline{A} \cup \overline{B}} \in \mathcal{B} \Rightarrow A \cap \mathcal{B} \in \mathcal{B}.$$

In summary, any combination of unions and intersections of sets belongs to the Borel algebra. In other words, operations of union or intersection of sets, or a combination of these operations, produce a set that belongs to the σ-algebra.

A.2 Basic Probability Theory

The first known published book on probability is *De Ludo Aleae* (About Games of Chance), by the Italian medical doctor and mathematician Girolamo Cardano (1501–1576), which came out in 1663, almost 90 years after his death. This book was a handbook for players, containing some discussion on probability.

The first mathematical treatise about the theory of probability, published in 1657, was written by the Dutch scientist Christian Huygens (1629–1695), a folder titled *De Ratiociniis in Ludo Aleae* (About Reasoning in Games of Chance).

Abraham de Moivre (1667–1754) was an important mathematician who worked on the development of probability theory. He wrote a book of great influence in his time, called *Doctrine of Chances*. The law of large numbers was discussed by Jacques Bernoulli (1654–1705), Swiss mathematician, in his work Ars Conjectandi (The Art of Conjecturing).

The study of probability was improved in the eighteenth and nineteenth centuries, being worth of mentioning the works of French mathematicians Pierre-Simon de Laplace (1749–1827) and Siméon Poisson (1781–1840) as well as the German mathematician Karl Friedrich Gauss (1777–1855).

A.2.1 The Axioms of Probability

The basic axioms of probability were established by Andrei Nikolaevich Kolmogorov (1903–1987) and allowed the development of the complete theory. The three statements are as follows (Papoulis, 1983):

1. Axiom 1 – $P(\Omega) = 1$, in which Ω denotes the sample space or universal set and $P(\cdot)$ denotes the associated probability;
2. Axiom 2 – $P(A) \geq 0$, in which A denotes an event belonging to the sample space;
3. Axiom 3 – $P(A \cup B) = P(A) + P(B)$, in which A and B are mutually exclusive events and $A \cup B$ denotes the union of events A and B.

Kolmogorov established a firm mathematical basis on which other theories rely, including the theory of stochastic processes, the communications theory, and the information theory, which use his axiomatic approach to probability
theory.

Kolmogorov's fundamental work was published in 1933, in Russian, and soon afterwards was translated to German with the title *Grundbegriffe der Wahrscheinlichkeits Rechnung* (Fundamentals of Probability Theory) (James, 1981). In this work, Kolmogorov managed to combine advanced set theory, developed by Cantor, with measure theory, established by Lebesgue, to produce what, to this date, is the modern approach to probability theory.

The application of the axioms makes it possible to deduce all results relative to probability theory. For example, the probability of the empty set,

$\emptyset = \{\}$, is easily calculated as follows. First, it is noticed that

$$\emptyset \cup \Omega = \Omega$$

since the sets $\emptyset$ and Ω are disjoint. Thus, it follows that

$$P(\emptyset \cup \Omega) = P(\Omega) = P(\emptyset) + P(\Omega) = 1 \Rightarrow P(\emptyset) = 0.$$

In the case of sets A and B, which are not disjoint, it follows that

$$P(A \cup B) = P(A) + P(B) - P(A \cap B).$$

A.2.2 Bayes' Rule

Bayes' rule, which is essential for the development of ,information theory, concerns the computation of conditional probabilities and can be expressed by the following definition:

$$P(A|B) = \frac{P(A \cap B)}{P(B)},$$

assuming $P(B) \neq 0$.

An equivalent manner of expressing the same result is the following:

$$P(A \cap B) = P(A|B) \cdot P(B) , \ P(B) \neq 0.$$

Some important properties of sets are presented next, in which A and B denote events from a given sample space.

- If A is independent of B, then $P(A|B) = P(A)$. It then follows that $P(B|A) = P(B)$ and that B is independent of A.
- If $B \subset A$, then: $P(A|B) = 1$.
- If $A \subset B$, then: $P(A|B) = \frac{P(A)}{P(B)} \geq P(A)$.
- If A and B are independent events, then $P(A \cap B) = P(A) \cdot P(B)$.
- If $P(A) = 0$ or $P(A) = 1$, then event A is independent of itself.
- If $P(B) = 0$, then $P(A|B)$ can assume any arbitrary value. Usually, in this case, one assumes $P(A|B) = P(A)$.
- If events A and B are disjoint, and non-empty, then they are dependent.

A partition is a possible splitting of the sample space into a family of subsets, in a manner that the subsets in this family are disjoint and their union coincides with the sample space. It follows that any set in the sample

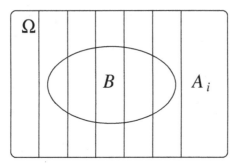

Figure A.5　Partition of set B by a family of sets $\{A_i\}$.

space can be expressed by using a partition of that sample space and, thus, be written as a union of disjoint events.

The following property can be illustrated by means of the Venn diagram, as illustrated in Figure A.5:

$$B = B \cap \Omega = B \cap \cup_{i=1}^{M} A_i = \cup_{i=1}^{N} B \cap A_i.$$

It now follows that

$$P(B) = P(\cup_{i=1}^{N} B \cap A_i) = \sum_{i=1}^{N} P(B \cap A_i),$$

$$P(A_i|B) = \frac{P(A_i \cap B)}{P(B)} = \frac{P(B|A_i) \cdot P(A_i)}{\sum_{i=1}^{N} P(B \cap A_i)} = \frac{P(B|A_i) \cdot P(A_i)}{\sum_{i=1}^{N} P(B|A_i) \cdot P(A_i)}.$$

A.3 Random Variables

A random variable (r.v.) X represents a mapping of the sample space on the line, that is, the set of real numbers. A random variable is usually characterized by a cumulative probability function (CPF) $P_X(x)$ or by a probability density function (pdf) $p_X(x)$.

Example: A random variable with a uniform probability density function, in the interval $[0,1]$, is described by the formula $p_X(x) = u(x) - u(x - 1)$. It follows, by Axiom 1, that

$$\int_{-\infty}^{+\infty} p_X(x)dx = 1. \tag{A.1}$$

In general, for a given probability distribution, the probability that X belongs to the interval $(a,b]$ is given by

$$P(a < x \le b) = \int_a^b p_X(x)dx. \tag{A.2}$$

The cumulative probability function $P_X(x)$, of a random variable X, is defined as the integral of $p_X(x)$,

$$P_X(x) = \int_{-\infty}^x p_X(t)dt. \tag{A.3}$$

A.3.1 Expected Value of a Random Variable

Let $f(X)$ denote a function of a random variable X. The average value, or expected value, of the function $f(X)$ with respect to X is defined as

$$E[f(X)] = \int_{-\infty}^{+\infty} f(x)p_X(x)dx. \tag{A.4}$$

The following properties of the expected value follow from eqn (A.4):

$$E[\alpha X] = \alpha E[X], \tag{A.5}$$

$$E[X + Y] = E[X] + E[Y], \tag{A.6}$$

and if X and Y are independent random variables, then

$$E[XY] = E[X] \cdot E[Y]. \tag{A.7}$$

A.3.2 Moments of a Random Variable

The kth moment of a random variable X is defined as

$$m_k = E[X^k] = \int_{-\infty}^{+\infty} x^k p_X(x)dx. \tag{A.8}$$

Various moments of X have special importance and physical interpretation, as defined in the following:

- $E[X]$, arithmetic mean, average value, average voltage, statistical mean;
- $E[X^2]$, quadratic mean, total power;
- $E[X^3]$, measure of asymmetry of the probability density function;
- $E[X^4]$, measure of flatness of the probability density function.

A.3.3 Variance of a Random Variable

The variance of a random variable X is an important quantity in communication theory, usually meaning AC power, defined as follows:

$$V[X] = \sigma_X^2 = E\left[(X - E[X])^2\right]. \tag{A.9}$$

The standard deviation σ_X is defined as the square root of the variance of X.

A.3.4 Characteristic Function

The characteristic function $P_X(w)$, also called moment generating function, of a random variable X is usually defined based on the Fourier transform of the pdf of X, which is equivalent to substitute $f(x) = e^{-jwx}$ in eqn (A.4), that is,

$$P_X(w) = E[e^{-jwx}] = \int_{-\infty}^{+\infty} e^{-jwx} p_X(x)dx, \text{ in which } j = \sqrt{-1}. \tag{A.10}$$

The statistical moments of a random variable X can also be obtained directly from the characteristic function, as follows:

$$m_i = \frac{1}{(-j)^i} \frac{\partial^i P_X(w)}{\partial w^i} \Big|_{w=0}. \tag{A.11}$$

Given that X is a random variable, it follows that $Y = f(X)$ is also a random variable, obtained by the application of the transformation $f(\cdot)$. The pdf of Y is related to that of X by the formula (Blake, 1987)

$$p_Y(y) = \frac{p_X(x)}{|dy/dx|} \Big|_{x=f^{-1}(y)}, \tag{A.12}$$

in which $f^{-1}(\cdot)$ denotes the inverse function of $f(\cdot)$. This formula assumes the existence of the inverse function of $f(\cdot)$ as well as its derivative in all points.

A.3.4.1 Two Important Distributions
1. **Gaussian random variable**
    The random variable X with pdf

$$p_X(x) = \frac{1}{\sigma_X \sqrt{2\pi}} e^{-\frac{(x-m_X)^2}{2\sigma_X^2}} \tag{A.13}$$

is called a Gaussian (or normal) random variable. The Gaussian random variable plays an extremely important role in engineering, considering that many well-known processes can be described or approximated by this pdf. The noise present in either analog or digital communications systems usually can be considered Gaussian as a consequence of the influence of many factors (Leon-Garcia, 1989). In eqn (A.13), m_X represents the average value and σ_X^2 represents the variance of X.

2. **Sinusoidal random variable**

A sinusoidal tone $X(t) = V \cos(\omega_0 t + \phi)$, in which V represents a constant amplitude, ω_0 is a constant frequency, and ϕ is a uniformly distributed random variable, has the following pdf:

$$p_X(x) = \frac{1}{\pi \sqrt{V^2 - x^2}}, \quad |x| < V. \tag{A.14}$$

A.3.5 Joint Random Variables

Consider that X and Y represent a pair of real random variables, with joint pdf $p_{XY}(x,y)$, as illustrated in Figure A.6. The probability of x and y being simultaneously in the region defined by the polygon [abcd] is given by the expression (Alencar, 2014b).

$$\text{Prob}(a < x < b, c < y < d) = \int_a^b \int_c^d p_{XY}(x,y)dxdy. \tag{A.15}$$

The individual pdf's of X and Y, also called marginal distributions, result from the integration of the joint pdf as follows:

$$p_X(x) = \int_{-\infty}^{+\infty} p_{XY}(x,y)dy, \tag{A.16}$$

and

$$p_Y(y) = \int_{-\infty}^{+\infty} p_{XY}(x,y)dx. \tag{A.17}$$

The joint average $E[f(X,Y)]$ is calculated as

$$E[f(X,Y)] = \int_{-\infty}^{+\infty} \int_{-\infty}^{+\infty} f(x,y)p_{XY}(x,y)dxdy, \tag{A.18}$$

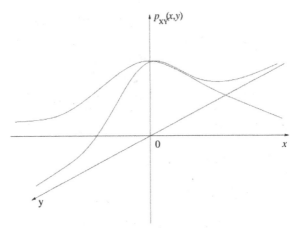

Figure A.6   Joint probability density function.

for an arbitrary function $f(X,Y)$ of X and Y.

The joint moments m_{ik}, of order ik, are calculated as

$$m_{ik} = E[X^i,Y^k] = \int_{-\infty}^{+\infty} \int_{-\infty}^{+\infty} x^i y^k p_{XY}(xy)dxdy. \qquad (A.19)$$

The two-dimensional characteristic function is defined as the two-dimensional Fourier transform of the joint probability density $p_{XY}(x,y)$

$$P_{XY}(\omega, \nu) = E[e^{-j\omega X - j\nu Y}]. \qquad (A.20)$$

When the sum $Z = X + Y$ of two statistically independent r.v.s is considered, it is noticed that the characteristic function of Z turns out to be

$$P_Z(\omega) = E[e^{-j\omega Z}] = E[e^{-j\omega(X+Y)}] = P_X(\omega) \cdot P_Y(\omega). \qquad (A.21)$$

As far as the pdf of Z is concerned, it can be said that

$$p_Z(z) = \int_{-\infty}^{\infty} p_X(\rho)p_Y(z - \rho)d\rho, \qquad (A.22)$$

or

$$p_Z(z) = \int_{-\infty}^{\infty} p_X(z - \rho)p_Y(\rho)d\rho. \qquad (A.23)$$

Equivalently, the sum of two statistically independent r.v.s has a pdf given by the convolution of the respective pdfs of the r.v.s involved in the sum.

The random variables X and Y are called uncorrelated if $E[XY] = E[X] \cdot E[Y]$. The criterion of statistical independence of random variables, which is stronger than correlation, is satisfied if $p_{XY}(x,y) = p_X(x) \cdot p_Y(y)$.

Appendix B

Cryptoalgorithms and Cryptosystems

> "Encryption works. Properly implemented strong crypto systems are one of the few things that you can rely on."

> Edward Snowden

Cryptographic algorithms are used for data encryption, authentication, and digital signatures, with the objective of linking cryptographic keys to machine or user identities.

The following list presents some cryptographic algorithms and cryptosystems that are proposed or employed for a variety of functions, including electronic commerce and secure communications (Kessler, 2021; Stallings, 1999).

AES – The Advanced Encryption Standard (AES) is a symmetric block cipher, developed by Vincent Rijmen and Joan Daemen, used by the United States government to protect classified information. The specification for the encryption of electronic data was adopted by the US National Institute of Standards and Technology (NIST), in 2001.

Bitmessage – A decentralized, encrypted, peer-to-peer, trustless communications protocol for message exchange. The decentralized design, outlined in "Bitmessage: A Peer-to-Peer Message Authentication and Delivery System," is conceptually based on the Bitcoin model.

Blowfish – A symmetric key block cipher that provides a good encryption rate in software, Blowfish has no effective cryptanalysis, as of now. Designed in 1993 by Bruce Schneier, Blowfish has a 64-bit block size and a variable key length from 32 bits up to 448 bits and has been included in many cipher suites and encryption products.

233

Capstone – This algorithm has been discontinued, but it was approved by the NIST and National Security Agency (NSA) project under the Bush Sr. and Clinton administrations for publicly available strong cryptography with keys escrowed by the government (NIST and the Treasury Dept.). Capstone included one or more tamper-proof computer chips for implementation (Clipper), a secret key encryption algorithm (Skipjack), digital signature algorithm (DSA), key exchange algorithm (KEA), and secure hash algorithm (SHA).

Challenge-Handshake Authentication Protocol (CHAP) – An authentication scheme that allows one party to prove who they are to a second party by demonstrating knowledge of a shared secret without actually divulging that shared secret to a third party who might be listening. It has been described in RFC 1994.

Chips-Message Robust Authentication (CHIMERA) – A scheme proposed for authenticating navigation data and the spreading code of civilian signals in the global positioning system (GPS). This is an anti-spoofing mechanism to protect the unencrypted civilian signals; GPS military signals are encrypted.

Clipper – The computer chip that would implement the Skipjack encryption scheme. The Clipper chip was to have had a deliberate backdoor so that material encrypted with this device would not be beyond the government's reach. Described in 1993, Clipper was dead by 1996.

Cryptography Research and Evaluation Committees (CRYPTEC) – Similar in concept to the NIST AES process and NESSIE, CRYPTEC is the Japanese government's process to evaluate algorithms submitted for government and industry applications. CRYPTEX maintains a list of public key and secret key ciphers, hash functions, MACs, and other crypto algorithms approved for various applications in government environments.

Data Encryption Standard (DES) – The DES is a symmetric key algorithm for the encryption of digital data. It first came into use in 1976 in the United States. DES is a block cipher based on symmetric key cryptography and uses a 56-bit key. DES was considered very secure for some time, but it is no longer considered a safe algorithm.

Derived Unique Key Per Transaction (DUKPT) – A key management scheme used for debit and credit card verification with point-of-sale (POS) transaction systems, automated teller machines (ATMs), and other financial applications. In DUKPT, a unique key is derived for each transaction based upon a fixed and shared key in such a way that knowledge of one derived key does not easily yield knowledge of other keys (including the fixed key). Therefore, if one of the derived keys is compromised, neither past nor subsequent transactions are endangered. DUKPT is specified in American National Standard (ANS); ANSI X9.24-1:2009 Retail Financial Services Symmetric Key Management Part 1: Using Symmetric Techniques) and can be purchased at the ANSI X9.24 Web page.

ECRYPT Stream Cipher Project (eSTREAM) – The eSTREAM project came about as a result of the failure of the NESSIE project to produce a stream cipher that survived cryptanalysis. It ran from 2004 to 2008 with the primary purpose of promoting the design of efficient and compact stream ciphers. As of September 2008, the eSTREAM suite contains seven ciphers.

Escrowed Encryption Standard (EES) – Largely unused, a controversial crypto scheme employing the SKIPJACK secret key crypto algorithm and a Law Enforcement Access Field (LEAF) creation method. LEAF was one part of the key escrow system and allowed for decryption of ciphertext messages that had been intercepted by law enforcement agencies. It is described more in FIPS PUB 185. It is archived and no longer in force.

Federal Information Processing Standards (FIPS) – These computer security and crypto-related FIPS PUBs are produced by the US National Institute of Standards and Technology (NIST) as standards for the US Government. Current FIPS related to cryptography include:

- FIPS PUB 140-2: Security Requirements for Cryptographic Modules.
- FIPS PUB 180-4: Secure Hash Standard (SHS).
- FIPS PUB 186-4: Digital Signature Standard (DSS).
- FIPS PUB 197: Advanced Encryption Standard (AES).
- FIPS PUB 198-1: The Keyed-Hash Message Authentication Code (HMAC).

- FIPS PUB 202: SHA-3 Standard: Permutation-Based Hash and Extendable-Output Functions.

Fortezza – A PCMCIA card developed by NSA that implements the Capstone algorithms, intended for use with the defense messaging service (DMS), originally called Tessera.

GOST – GOST is a family of algorithms defined in the Russian cryptographic standards. Although most of the specifications are written in Russian, a series of RFCs describe some of the aspects so that the algorithms can be used effectively in Internet applications:

- RFC 4357: Additional Cryptographic Algorithms for Use with GOST 28147-89, GOST R 34.10-94, GOST R 34.10-2001, and GOST R 34.11-94 Algorithms.
- RFC 4490: Using the GOST 28147-89, GOST R 34.11-94, GOST R 34.10-94, and GOST R 34.10-2001 Algorithms with Cryptographic Message Syntax (CMS).
- RFC 4491: Using the GOST R 34.10-94, GOST R 34.10-2001, and GOST R 34.11-94 Algorithms with the Internet X.509 Public Key Infrastructure Certificate and CRL Profile.
- RFC 5830: GOST 28147-89: Encryption, Decryption, and Message Authentication Code (MAC) Algorithms.
- RFC 6986: GOST R 34.11-2012: Hash Function Algorithm.
- RFC 7091: GOST R 34.10-2012: Digital Signature Algorithm (Updates RFC 5832: GOST R 34.10-2001).
- RFC 7801: GOST R 34.12-2015: Block Cipher Kuznyechik.
- RFC 7836: Guidelines on the Cryptographic Algorithms to Accompany the Usage of Standards GOST R 34.10-2012 and GOST R 34.11-2012.
- RFC 8891: GOST R 34.12-2015: Block Cipher Magma.

Hash – A cryptographic hash function is a specific family of algorithms which is used to guarantee the integrity of data transmission. This function creates a hash value of a fixed length and with a certain number of properties for each message. The hash value produced by the MD5 algorithm have 128 bits, the algorithm SHA-1 produces a hash of 160 bits, and SHA-256 produces 256-bit hashes.

IP Security (IPsec) – The IPsec protocol suite is used to provide privacy and authentication services at the IP layer. An overview of the protocol suite and of the documents comprising IPsec can be found in RFC 2411. Other documents include:

- RFC 4301: IP security architecture.
- RFC 4302: IP authentication header (AH), one of the two primary IPsec functions; AH provides connectionless integrity and data origin authentication for IP datagrams and protects against replay attacks.
- RFC 4303: IP encapsulating security payload (ESP), the other primary IPsec function; ESP provides a variety of security services within IPsec.
- RFC 4304: Extended sequence number (ESN) addendum allows for negotiation of a 32- or 64-bit sequence number, used to detect replay attacks.
- RFC 4305: Cryptographic algorithm implementation requirements for ESP and AH.
- RFC 5996: The Internet Key Exchange (IKE) protocol, version 2, providing for mutual authentication and establishing and maintaining security associations.
- IKE v1 was described in three separate documents, RFC 2407 (application of ISAKMP to IPsec), RFC 2408 (ISAKMP, a framework for key management and security associations), and RFC 2409 (IKE, using part of Oakley and part of SKEME in conjunction with ISAKMP to obtain authenticated keying material for use with ISAKMP and for other security associations such as AH and ESP). IKE v1 is obsoleted with the introduction of IKEv2.
- RFC 4307: Cryptographic algorithms used with IKEv2.
- RFC 4308: Crypto suites for IPsec, IKE, and IKEv2.
- RFC 4309: The use of AES in CBC-MAC mode with IPsec ESP.
- RFC 4312: The use of the Camellia cipher algorithm in IPsec.
- RFC 4359: The use of RSA/SHA-1 signatures within ESP and AH.
- RFC 4434: Describes AES-XCBC-PRF-128, a pseudo-random function derived from the AES for use with IKE.
- RFC 2403: Describes the use of the HMAC with MD5 algorithm for data origin authentication and integrity protection in both AH and ESP.

- RFC 2405: Describes the use of DES-CBC (DES in cipher block chaining mode) for confidentiality in ESP.
- RFC 2410: Defines use of the NULL encryption algorithm (i.e., provides authentication and integrity without confidentiality) in ESP.
- RFC 2412: Describes OAKLEY, a key determination and distribution protocol.
- RFC 2451: Describes use of cipher block chaining (CBC) mode cipher algorithms with ESP.
- RFCs 2522 and 2523: Description of Photuris, a session-key management protocol for IPsec.

In addition, RFC 6379 describes Suite B cryptographic suites for IPsec and RFC 6380 describes the Suite B profile for IPsec.

IPsec was first proposed for use with IP version 6 (IPv6) but can also be employed with the current IP version, IPv4.

Internet Security Association and Key Management Protocol (ISAKMP/OAKLEY) – The protocol ISAKMP/OAKLEY provides an infrastructure for Internet secure communications. ISAKMP, designed by the National Security Agency (NSA) and described in RFC 2408, is a framework for key management and security associations, independent of the key generation and cryptographic algorithms actually employed. The OAKLEY Key Determination Protocol, described in RFC 2412, is a key determination and distribution protocol using a variation of Diffie–Hellman.

Kerberos – A secret key encryption and authentication system, designed to authenticate requests for network resources within a user domain rather than to authenticate messages. Kerberos also uses a trusted third-party approach; a client communications with the Kerberos server to obtain credentials so that it may access services at the application server. Kerberos V4 used DES to generate keys and encrypt messages; Kerberos V5 uses DES and other schemes for key generation.

Microsoft added support for Kerberos V5, with some proprietary extensions, in Windows 2000 Active Directory. There are many Kerberos articles posted at Microsoft's Knowledge Base, notably "Kerberos Explained."

Keyed-Hash Message Authentication Code (HMAC) – A message authentication scheme based on secret key cryptography and the secret key shared between two parties rather than public key methods, described in FIPS PUB 198 and RFC 2104. (See Section 5.19 for details on HMAC operation.)

Message Digest (MD) – Hashing is a technique in which an algorithm, or hash function, is applied to a portion of data to create a unique digital fingerprint, that is, a fixed-size parameter. If a hacker changes the data, the hash function generates a different output, called the hash value, and the receiver will know that the data has been changed. The algorithms MD2, MD4, and MD5 use a message digest, the hash value, that is 128 bits in length. They were created by Ron Rivest and are popularly used for digital signatures.

Message Digest Cipher (MDC) – Invented by Peter Gutman, MDC turns a one-way hash function into a block cipher.

MIME Object Security Services (MOSS) – Designed as a successor to PEM to provide PEM-based security services to MIME messages, which is described in RFC 1848. It was never widely implemented and is now defunct.

Mujahedeen Secret – A Windows GUI, PGP-like cryptosystem. Developed by supporters of Al-Qaeda, the program employs the five finalist AES algorithms, namely, MARS, RC6, Rijndael, Serpent, and Twofish. Also described in Inspire Magazine, Issue 1, pp. 41-44 and Inspire Magazine, Issue 2, pp. 58–59.

New European Schemes for Signatures, Integrity, and Encryption (NESSIE) – NESSIE was an independent project meant to augment the work of NIST during the AES adoption process by putting out an open call for new cryptographic primitives. The NESSIE project ran from about 2000–2003. While several new block cipher, PKC, MAC, and digital signature algorithms were found during the NESSIE process, no new stream cipher survived cryptanalysis. As a result, the ECRYPT Stream Cipher Project (eSTREAM) was created.

NSA Suite B Cryptography – An NSA standard for securing information at the SECRET level. The standard proposes:

- Advanced encryption standard (AES), with key sizes of 128 and 256 bits, per FIPS PUB 197 for encryption.
- The ephemeral unified model and the one-pass Diffie–Hellman, also referred to as ECDH, uses the curves with 256- and 384-bit prime *moduli*, per NIST Special Publication 800-56A for key exchange.
- Elliptic curve digital signature algorithm (ECDSA), which uses the curves with 256- and 384-bit prime *moduli*, per FIPS PUB 186-3 for digital signatures.
- Secure hash algorithm (SHA) uses 256 and 384 bits per FIPS PUB 180-3 for hashing. There are several variations on this algorithm, including SHA1, SHA256, SHA384, and SHA512. The length of the hash value is the main difference between them.

RFC 6239 describes Suite B cryptographic suites for secure shell (SSH) and RFC 6379 describes Suite B cryptographic suites for secure IP (IPsec).

RFC 8423 reclassifies the RFCs related to the Suite B cryptographic algorithms as historic, and it discusses the reasons for doing so.

Pretty Good Privacy (PGP) – A family of cryptographic routines for hboxe-mail, file, and disk encryption developed by Philip Zimmermann. PGP 2.6.x uses RSA for key management and digital signatures, IDEA for message encryption, and MD5 for computing the message's hash value. PGP 5.x, formerly known as PGP 3, uses Diffie–Hellman/DSS for key management and digital signatures; IDEA, CAST, or 3DES for message encryption; and MD5 or SHA for computing the message's hash value. OpenPGP, described in RFC 2440, is an open definition of security software based on PGP 5.x. The GNU privacy guard (GPG) is a free software version of OpenPGP.

Privacy Enhanced Mail (PEM) – An IETF standard for secure transmission of electronic mail over the Internet, including provisions for encryption (DES), authentication, and key management (DES and RSA). It is developed by the IETF but never widely used and described in the following RFCs:

- RFC 1421: Part I, Message Encryption and Authentication Procedures.

- RFC 1422: Part II, Certificate-Based Key Management.
- RFC 1423: Part III, Algorithms, Modes, and Identifiers.
- RFC 1424: Part IV, Key Certification and Related Services.

Private Communication Technology (PCT) – Developed by Microsoft for secure communication on the Internet. PCT supported Diffie–Hellman, Fortezza, and RSA for key establishment; DES, RC2, RC4, and triple-DES for encryption; and DSA and RSA for message signatures. It was never widely used and ended up superseded by SSL and TLS.

Public Key Infrastructure (PKI) – Asymmetric cryptography is the basic technology behind public key infrastructure (PKI), which permits scalable issuance, revocation, and management of digital certificates.

Secure Electronic Transaction (SET) – A communications protocol for securing credit card transactions, developed by MasterCard and VISA, in cooperation with IBM, Microsoft, RSA, and other companies. Merged two other protocols: Secure Electronic Payment Protocol (SEPP), an open specification for secure bank card transactions over the Internet developed by CyberCash, GTE, IBM, MasterCard, and Netscape; and secure transaction Technology (STT), a secure payment protocol developed by Microsoft and Visa International. It supports DES and RC4 for encryption, and RSA for signatures, key exchange, and public key encryption of bank card numbers. SET V1.0 is described in Book 1, Book 2, and Book 3. SET has been superseded by SSL and TLS.

Secure HyperText Transfer Protocol (S-HTTP) – An extension to HTTP to provide secure exchange of documents over the World Wide Web. Supported algorithms include RSA and Kerberos for key exchange, and DES, IDEA, RC2, and Triple-DES for encryption. It is described in RFC 2660. The protocol S-HTTP did not have the same success as HTTP over SSL (https).

Secure Multipurpose Internet Mail Extensions (S/MIME) – An IETF secure e-mail scheme superseding PEM and adding digital signature and encryption capability to Internet MIME messages. S/MIME Version 3.1 is described in RFCs 3850 and 3851, and employs the cryptographic message syntax described in RFCs 3369 and 3370.

Secure Sockets Layer (SSL) – Developed, in 1995, by Netscape Communications to provide application-independent security and privacy over the Internet. SSL is designed so that protocols such as HTTP, FTP (File Transfer Protocol), and Telnet can operate over it transparently. SSL allows both server authentication (mandatory) and client authentication (optional). RSA is used during negotiation to exchange keys and identify the actual cryptographic algorithm (DES, IDEA, RC2, RC4, or 3DES) to use for the session. SSL also uses MD5 for message digests and X.509 public key certificates.

The SSL was found to be breakable soon after the IETF announced that formation of group to work on TLS and RFC 6176 specifically prohibits the use of SSL v2.0 by TLS clients. SSL version 3.0 is described in RFC 6101. All versions of SSL are now deprecated in favor of TLS; TLS v1.0 is sometimes referred to as SSL v3.1.

Server Gated Cryptography (SGC) – Microsoft extension to SSL that provided strong encryption for online banking and other financial applications using RC2 (128-bit key), RC4 (128-bit key), DES (56-bit key), or 3DES (equivalent of 168-bit key). Use of SGC required a Windows NT Server running Internet Information Server (IIS) 4.0 with a valid SGC certificate. SGC was available in 32-bit Windows versions of Internet Explorer (IE) 4.0; support for Mac, Unix, and 16-bit Windows versions of IE was planned, but never materialized, and SGC was made moot when browsers started to ship with 128-bit encryption.

ShangMi (SM) Cipher Suites – A suite of authentication, encryption, and hash algorithms from the People's Republic of China.

- SM2 Cryptography Algorithm: A public key crypto scheme based on elliptic curves. An overview of the specification, in Chinese, can be found in GM/T 0009-2012. Additional specifications can be found in:
    - GB/T 32918.1-2016, Part 1: General.
    - GB/T 32918.2-2016, Part 2: Digital signature algorithm.
    - GB/T 32918.3-2016, Part 3: Key exchange protocol.
    - GB/T 32918.4-2016, Part 4: Public key encryption algorithm.
    - GB/T 32918.5-2017, Part 5: Parameter definition.
- SM3 Cryptographic Hash Algorithm: A hash algorithm operating on 512-bit blocks to produce a 256-bit hash value. It is described in GB/T 32905-2016.

- SM4 Block Cipher Algorithm: A Feistel block cipher algorithm with a block length and key length of 128 bits and 32 rounds. It is described in GB/T 32907-2016.
- An application of the ShangMi Cipher Suites in TLS can be found in RFC 8998.

Signal Protocol – A protocol for providing end-to-end encryption for voice calls, video calls, and instant messaging, including group chats. Employing a combination of AES, ECC, and HMAC algorithms, it offers such features as confidentiality, integrity, authentication, forward/future secrecy, and message repudiation. Signal is particularly interesting because of its lineage and widespread use. The Signal Protocol's earliest versions were known as TextSecure, first developed by Open Whisper Systems in 2013.

TextSecure itself was based on a 2004 protocol called Off-the-Record (OTR) Messaging, designed as an improvement over OpenPGP and S/MIME. TextSecure v2 (2014) introduced a scheme called the Axolotl Ratchet for key exchange and added additional communication features. After subsequent iterations improving key management (and the renaming of the key exchange protocol to Double Ratchet), additional cryptographic primitives, and the addition of an encrypted voice calling application (RedPhone), TextSecure was renamed Signal Protocol in 2016.

The Ratchet key exchange algorithm is at the heart of the power of this system. Most messaging apps employ the users' public and private keys; the weakness here is that if the phone falls into someone else's hands, all of the messages on the device, including deleted messages, can be decrypted. The Ratchet algorithm generates a set of temporary keys for each user, based upon that user's public/private key pair. When two users exchange messages, the Signal Protocol creates a secret key by combining the temporary and permanent pairs of public and private keys for both users. Each message is assigned its own secret key. Because the generation of the secret key requires access to both users' private keys, it exists only on their two devices. The Signal Protocol has been employed in:

- WhatsApp (introduced 2014).
- G Data Software's Secure Chat (introduced in 2015; service discontinued in 2018).

- Google's Allo application was introduced in 2016 and discontinued in favor of Messages application in 2019.
- Facebook Messenger was introduced in 2016.
- Skype's Private Conversations mode was introduced in 2018.
- All of Google's Rich Communication Services (RCS) on Android systems was introduced in 2020.

A reasonably good write-up of the protocol can be found in "Demystifying the Signal Protocol for End-to-End Encryption (E2EE)" by Kozhukhovskaya, Mora, and Wong (2017).

Simple Authentication and Security Layer (SASL) – A framework for providing authentication and data security services in connection-oriented protocols (such as TCP), described in RFC 4422. It provides a structured interface and allows new protocols to reuse existing authentication mechanisms and allows old protocols to make use of new mechanisms.

It has been common practice on the Internet to permit anonymous access to various services, employing a plain-text password using a username of anonymous and a password of an email address or some other identifying information. New IETF protocols disallow plain-text logins. The Anonymous SASL Mechanism (RFC 4505) provides a method for anonymous logins within the SASL framework.

Simple Key-Management for Internet Protocol (SKIP) – Key management scheme for secure IP communication, specifically for IPsec, and designed by Aziz and Diffie. SKIP essentially defines a public key infrastructure for the Internet and even uses X.509 certificates. Most public key cryptosystems assign keys on a per-session basis, which is inconvenient for the Internet since IP is connectionless. Instead, SKIP provides a basis for secure communication between any pair of Internet hosts. SKIP can employ DES, 3DES, IDEA, RC2, RC5, MD5, and SHA-1. As it happened, SKIP was not adopted for IPsec; IKE was selected instead.

SM9 – Chinese Standard GM/T0044-2016 SM9 (2016) is the Chinese national standard for identity-based cryptography. SM9 comprises three cryptographic algorithms, namely the identity based digital signature algorithm, identity based key agreement algorithm, and identity based key encapsulation algorithm (allowing one party to securely send a symmetric key to another party). The SM9 scheme is also described in The SM9 Cryptographic Schemes (Z. Cheng).

Telegram – Telegram, launched in 2013, is a cloud-based instant messaging and voice over IP (VoIP) service, with client app software available for all major computer and mobile device operating systems. Telegram allows users to exchange messages, photos, and videos, for example, and supplies end-to-end encryption using a protocol called MTProto. stickers, audio and files of any type. MTProto employs 256-bit AES, 2048-bit RSA, and Diffie–Hellman key exchange. There have been several controversies with Telegram, not the least of which has to do with the nationality of the founders and the true location of the business as well as some operation issues.

From a cryptological viewpoint, however, a description can be found in "On the CCA (in)security of MTProto" (Jakobsen & Orlandi, 2015), who describes some of the crypto weaknesses of the protocol; specifically, that "MTProto does not satisfy the definitions of authenticated encryption (AE) or indistinguishably under chosen-ciphertext attack (IND-CCA)."

Transmission Control Protocol (TCP) encryption (tcpcrypt) – Before 2019, the major part of Internet TCP traffic was not encrypted. The two primary reasons for this were the legacy protocols, that have no mechanism with which to employ encryption, and several legacy applications that cannot be upgraded to allow new encryption to be added. The response from the IETF's TCP Increased Security Working Group was to define a transparent way within the transport layer (TCP) with which to invoke encryption. The TCP Encryption Negotiation Option (TCP-ENO) addresses these two problems with an out-of-band, fully backward-compatible TCP option with which to negotiate the use of encryption. TCP-ENO is described in RFC 8547 and tcpcrypt, an sencryption protocol to protect TCP streams, is described in RFC 8548.

Transport Layer Security (TLS) – TLS v1.0 is an IETF specification (RFC 2246) intended to replace SSL v3.0. TLS v1.0 employs Triple-DES (secret key cryptography), SHA (hash), Diffie–Hellman (key exchange), and DSS (digital signatures). TLS v1.0 was vulnerable to attack and updated by v1.1 (RFC 4346), which is now classified as an HISTORIC specification. TLS v1.1 was replaced by TLS v1.2 (RFC 5246) and, subsequently, by v1.3 (RFC 8446).

The TLS protocol is designed to operate over TCP. The IETF developed the Datagram Transport Layer Security (DTLS) protocol to operate over UDP. DTLS v1.2 is described in RFC 6347.

TrueCrypt – Open source, multi-platform cryptography software that can be used to encrypt a file, partition, or entire disk. One of TrueCrypt's more interesting features is that of plausible deniability with hidden volumes or hidden operating systems. The original web site, truecrypt.org, suddenly went dark in May 2014. The current fork of TrueCrypt is VeraCrypt.

X.509 – ITU-T recommendation for the format of certificates for the public key infrastructure. Certificates map (bind) a user identity to a public key. The IETF application of X.509 certificates is documented in RFC 5280. An Internet X.509 public key infrastructure is further defined in RFC 4210 (Certificate Management Protocols) and RFC 3647 (Certificate Policy and Certification Practices Framework).

Appendix C

Glossary of Information Security Terms

> "People often represent the weakest link in the security chain and are chronically responsible for the failure of security systems."
>
> Bruce Schneier

This glossary presents the fundamental definitions for understanding network security and information and includes the acronyms mostly used by professionals computer networks (Stallings, 1999; Alencar, 1999; Furiati, 1998; Design, 2001; Skycell, 2001; MobileWord, 2001; TIAB2B.com, 2001; Alencar, 2002, 2001). It is also based on the parameters established by the Center for Studies, Response and Security Incident Handling in Brazil (CERT.br), which is part of the Management Committee of Internet in Brazil (CERT.br, 2019).

Acceptable use policy – Also called a "Term of Use" or "Term of Service." Policy in which the rules for the use of computational resources are defined, the rights and responsibilities of those who use them, and uses and situations that are considered abusive.

Adware – It is an advertising software. specific type of spyware. Program is designed specifically to display advertisements. It can be used legitimately, when incorporated into programs and services, as a form of sponsorship or financial return for those who develop free programs or provide free services. It can also be used for malicious purposes when the advertisements shown are targeted, according to the user's navigation and without the user knowing that such monitoring is being carried out.

ANSI – American National Standards Institute. US-based organization that, among other activities, develops standards for telecommunications.

Antimalware – Tool that seeks to detect and then cancel or remove malicious code from a computer. Antivirus, antispyware, antirootkit, and antitrojan programs are examples of antimalware tools.

Anti-spam filter – Program that allows you to separate emails according to predefined rules. Used both for the management of mailboxes and for the selection of valid e-mails from the various spam received.

Antivirus – Type of antimalware tool developed to detect, cancel, and eliminate viruses and other types of malicious code from a computer. It can also include personal firewall functionality.

AP – Access point. Device that acts as a bridge between a wireless network and a traditional network.

ARP – Address Resolution Protocol. Enables the packaging of IP data into Ethernet packets, being used to find an Ethernet network through a specific IP number.

ARQ – Automatic ReQuest for retransmission or automatic request for retransmission. Elementary protection scheme against transmission errors in which the receiver detects errors and automatically requests the retransmission of information.

Artifact – Any information left by an attacker on a compromised system, such as programs, scripts, tools, logs, and files.

ASCII – American Standard Code for Information Interchange or code for information exchange. Each ASCII character is formed by a 7-bit code plus a parity bit.

ATM – Asynchronous transfer mode. Data transfer pattern using 53 bytes packets, with 5 bytes for addressing and 48 bytes of information.

Attack – Any attempt, successful or unsuccessful, to access or unauthorized use of a service, computer, or network.

Attacker – Person responsible for an attack, an attempted access, or unauthorized use of a program/computer.

AT&T – American Telephone and Telegraph. Long distance operator in the USA.

Auction and sale site for products – Type of site that mediates the purchase and sale of goods among users. Some sites of this type offer a management system whereby payment made by the buyer is only released to the seller when confirmation that the goods have been correctly received is sent.

AWGN – Additive white Gaussian noise or channel. Channel with broadband noise with flat power spectral density.

Backdoor – Type of malicious code. A program that allows an attacker to return to a compromised computer by including services created or modified for this purpose. Usually, this program is placed in such a way that it is not noticed.

Bandwidth – Amount of data that can be transmitted on a communication channel in a given time interval.

BCH – Bose–Chaudhuri–Hocquenghem. Family of linear block codes used for error correction.

BER – Bit error rate. Ratio between the number of bits received in error and the total number of bits transmitted. It is also known as error probability.

BGP Border Gateway Protocol – When two systems use BGP, they establish a TCP connection and then share their BGP tables. BGP uses a distance vectoring method and detects failures, sending activity messages to its neighbors at 30-second intervals. The protocol exchanges information about available networks with other BGP systems, including the full path of the systems that are between them.

Blacklist – List of e-mails, domains, or IP addresses, known sources of spam. Resource used both on servers and in e-mail reader programs to block messages suspected of being spam.

Bluetooth – Standard for data and voice communication technology based on radio frequency and intended for the connection of devices over short distances, allowing the formation of personal wireless networks.

BOOTP The Bootstrap protocol assigns an IP address to computers that do not have a hard disk. The server provides a file with the operating system to run it.

Botnet – Network formed by hundreds or thousands of computers infected with bots. It allows to enhance the harmful actions performed by bots and be used in denial of service attacks, fraud schemes, and sending spam, among others.

Bot – Type of malicious code. Program that, in addition to including functionality of worms, has communication mechanisms with the attacker that allow it to be controlled remotely. The process of infection and propagation of the bot is similar to that of the worm, that is, the bot is able to propagate automatically, exploiting existing vulnerabilities in programs installed on computers.

Broadband – Type of connection to the network with capacity above that achieved, usually in dial-up connection via telephone system. There is no definition of broadband metrics that is accepted by everyone, but it is common for broadband connections to be permanent and not switched, such as dial-up connections.

Brute force – Type of attack that consists of guessing, by trial and error, a username and password and, thus, executing processes and accessing websites, computers, and services in the name and with the same privileges of that user.

CCITT – Former name of ITU-R, ITU committee responsible for technical recommendations for telephone and data transmission systems.

CD – Compact disk.

Certification authority – Entity responsible for issuing and managing digital certificates. These certificates can be issued to different types of entities, such as: person, computer, department of an institution, institution, etc.

CODEC – COder/DECoder. Device that converts analog signals into digital signals and vice versa.

Code signing – A certificate for software developers to digitally sign the source code of applications created by them.

Collective shopping website – Type of website through which advertisers offer products, usually with big discounts, for a very short time and with limited quantities. It is considered as an intermediary between the companies that make the ads and the customers who buy the products.

Cookie – Small file that is saved on the computer when the user accesses a website and sent back to the same website when accessed again. It is used to maintain information about the user, such as shopping cart, product list, and browsing preferences.

COST – European Commission for Cooperation in Technical and Scientific Research.

CRC – Cyclic redundancy check. Method for detecting errors in serial transmissions.

Cryptoanalysis – The process of identifying or tracking failures or loopholes in a cryptographic system.

Cryptographic algorithm – The computational procedure used to encrypt and decrypt messages.

Cryptography – Science and art of writing messages in encrypted form or in code. It is part of a field of study that deals with secret communications. It is used, among other purposes, to: authenticate the identity of users; authenticate bank transactions; protect the integrity of electronic funds transfers; and protect the confidentiality of personal and commercial communications.

Cryptology – The study of cryptography and cryptanalysis.

CSLIP – Compressed SLIP is the data compression for the SLIP protocol, which uses the Van Jacobson algorithm to reduce the packet size. It is also used with PPP, being called CPPP.

CSR Key – Text file containing information for requesting the digital certificate and which is generated by the web server.

DDS – Distributed denial of service. Denial attack that occurs when a set of computers tries to take service out of other computers, networks, or programs.

Decryption – The process of converting an encrypted message to the original intelligible message.

Denial of service – Malicious activity by which an attacker uses a computer or mobile device to take a service, computer, or network connected to the Internet out of operation.

DHCP – Dynamic Host Configuration Protocol – Designates IP addresses for computers connected to the network. It is a server-based service that automatically assigns IP addresses to each computer that joins the network. DHCP can perform all BOOTP functions.

Dial-up connection – Switched connection to the Internet, made through an analog modem and a fixed telephone network line, which requires the modem to dial a phone number to access it.

Digital certificate – Electronic record consisting of a set of data that distinguishes an entity and associates it with a public key. It can be issued to people, companies, equipment, or services on the network (for example, a website) and can be approved for different uses, such as confidentiality and digital signature.

Digital signature – Code used to prove the authenticity and integrity of an information, that is, that it was actually generated by those who claim to have done so and that it has not been altered.

Distributed denial of service – Malicious, coordinated, and distributed activity by which a set of computers or mobile devices is used to shut down a service, a computer, or a network connected to the Internet.

DNS – Domain Name System, responsible for the translation, among other types, of machine/domain names to the corresponding IP address and vice versa.

DS-SS – Direct sequence spread spectrum or spectral spread by a direct sequence.

Eb/No – Energy-per-bit to noise power density ratio. Figure of merit for digital communication systems.

E-commerce – Any form of commercial transaction where the parties interact electronically. Set of computational techniques and technologies used to facilitate and execute commercial transactions for goods and services through the Internet.

E-commerce scam – Type of fraud in which a scammer, in order to obtain financial advantages, exploits the relationship of trust between the parties involved in a commercial transaction.

EGP – Exterior Gateway Protocol. Protocol used between routers of different systems.

E-mail spoofing – Counterfeiting of e-mail.

E-mail spoofing – Technique that consists of altering the header fields of an e-mail in order to make it appear that it was sent from a certain source when, in fact, it was sent from another source.

Encryption – Conversion of an intelligible message, clear text, to an encrypted message.

Ethernet – It is an interconnection architecture for local networks, a set of layers that provides the encapsulation of frames, before being sent to the network.

ETSI – European Telecommunications Standards Institute.

EV SSL – Extended validation secure socket layer. Advanced or extended validation SSL certificate.

Exploit – Malicious program designed to exploit a computer or program's vulnerability.

FCC – Federal Communications Commission. US government agency responsible for regulatory issues in the telecommunications area.

FEC – Forward error correction. Technique in which data is modified to increase its robustness to channel errors, with the aim of improving the probability of system error.

FFT – Fast Fourier transform.

Firewall – Security device used to divide and control access between computer networks.

Fixed broadband – Type of broadband connection that allows a computer to be connected to the Internet for long periods and with a low frequency of changing IP addresses.

Foursquare – Social network based on geolocation that, like other networks of the same type, uses the data provided by the GPS of the user's computer or mobile device to record the places where he passes.

FTP – File Transfer Protocol – Allows the transfer of files between two computers using a login and password.

GnuPG – Set of free and open source programs, which implements symmetric key encryption, asymmetric keys, and digital signature.

Greylisting – Filtering method for spam, implemented directly on the e-mail server, which temporarily refuses an e-mail and receives it only when it is resent. Legitimate e-mail servers, which behave correctly and in accordance with the specifications of the protocols, always forward messages. This method assumes that spammers rarely use legitimate servers and, therefore, do not resend their messages.

Harvesting – Technique used by spammers, which consists of scanning web pages and mailing list files, among others, in search of e-mail addresses.

Hash function – Cryptographic method that, when applied to information, regardless of its size, generates a single, fixed-size result, called a hash.

Hoax – Message that has alarming or false content and that usually has as its sender, or points as the author, some institution, important company, or government agency. Through a thorough reading of its content, it is usually possible to identify meaningless information and attempted strikes, such as chains and pyramids.

Hot fix – Security fix.

HTML – HyperText Markup Language is a protocol used for transferring web pages between the server and the client. It works as a kind of universal language that is used to create pages on the Internet.

HTPPS – Specifies the usefulness of HTTP through a security mechanism.

HTTP HyperText Transfer Protocol is used to transport HTML pages from network servers to browsers. The protocol performs communication between servers and browsers installed on client computers.

HTTP – in HyperText Transfer Protocol. Protocol used to transfer web pages between a server and a client (for example, the browser).

HTTPS – HyperText Transfer Protocol Secure over SSL. Protocol that combines the use of HTTP with security mechanisms, such as SSL and TLS, in order to provide secure connections.

ICMP – Internet Control Message Protocol (ICMP) manages and produces the error report for transporting data during communication between computers.

Identity theft – An act by which a person tries to impersonate another person, by assigning himself or herself a false identity, in order to obtain undue advantages. Some cases of identity theft can be considered a crime against the public faith, typified as false identity.

IDS – Intrusion detection system. Program, or a set of programs, whose function is to detect malicious or anomalous activities.

IDS – Intrusion detection system. Program with the function of detecting unauthorized activities on the computer.

IEEE – Institute of Electrical and Electronic Engineers. US organization responsible for publications and standards in several areas of electrical engineering.

IFFT – Inverse fast Fourier transform.

IGMP – Internet Group Management Protocol is used to support multicasting messages and to track groups of users on the computer network.

IGP – Interior Gateway Protocol – Each system on the Internet can choose its own routing protocol. RIP and OSPF are examples of IGP protocols.

IMAP4 – Internet Mail Access Protocol version 4 – Protocol that replaces POP3.

IN – Intelligent network. Structure built on the telephone network for the provision of additional services such as virtual private network, personal DDG, service 0800, 0900, connection with the Internet and a single number for users.

Inmarsat – International Maritime Satellite Corporation.

Invader – Person responsible for carrying out an invasion (compromise).

Invasion – Successful attack that results in accessing, manipulating, or destroying information on a computer.

IP address – Sequence of numbers associated with each computer connected to the Internet. In the case of IPv4, the IP address is divided into four groups, separated by periods and made up of numbers between 0 and 255, for example, "193.0.3.3." In the case of IPv6, the IP address is divided into up to eight groups, separated by colons and composed of hexadecimal numbers (numbers and letters from A to F) between 0 and FFFF, for example, "2019:DB9:C011:901D:DA28:126B::3."

IP – Internet Protocol. Except for ARP and RARP, all data from all protocols is packaged in an IP data packet, which addresses and manages data packets sent between computers.

ISDN – Integrated services digital network or integrated services digital network. Digital telephone service with speed between 57.6 and 128 kbit/s.

ISI – Intersymbol interference. Problem in digital communication system in which adjacent symbols in a sequence are distorted due to non-ideal frequency response, causing dispersion that interferes with the time domain in neighboring symbols.

IS – Interim standard.

ITU – International Telecommunication Union. Organization associated with the UN that regulates telecommunications worldwide .

ITU-R – ITU Radio Standardization Sector.

ITU-T – ITU Telecommunications Standards Sector.

JPEG – Joint Photographic Experts Group. Organization that, among other things, sets standards for images.

kbit/s – kilobits per second. Thousands of bits per second. It is recommended to use the shannon [Sh] unit, corresponding to bits per second (bits/s).

Keylogger – Specific type of spyware. Program capable of capturing and storing the keys typed by the user on the computer keyboard. Usually the activation of the keylogger is conditioned to a previous action of the user, such as access to a specific e-commerce site or Internet banking.

kHz – Kilohertz. Frequency unit that equates to 10^3 Hz.

LAN – Local area network or local computer network.

LBG – Linde–Buzo–Gray. Algorithm proposed by Linde, Buzo, and Gray in 1980, constituting the most widely used technique for dictionary design (vector quantization).

LMDS – Local multipoint distribution system. Fixed wireless system used for voice and data transmission.

Log – Activity log generated by programs and services on a computer. Technical term that refers to the registration of activities of various types, such as, for example, connection (information about the connection of a computer to the Internet) and access to applications (information about accessing a computer to an Internet application).

Malicious code – Generic term used to refer to programs designed to perform harmful actions and malicious activities on a computer or mobile device. Specific types of malicious code are: viruses, worm, bot, spyware, backdoor, Trojan horse, and rootkit.

Malvertising – Malicious advertising. Type of scam that consists of creating malicious ads and, through advertising services, displaying them on various web pages. Generally, the advertising service is tricked into believing that it is a legitimate ad and, by accepting it, it intermediates the presentation and makes it appear on several pages.

Malware – Malicious software. Malicious program or code.

Master key – Unique password used to protect (encrypt) other passwords.

Master password – Master key.

Mbit/s – Mega-bits per second. Millions of bits per second. it is recommended to use the unit Sh (bit per second).

MHz – Megahertz. Frequency unit that equates to 10^6 Hz.

MIPS – Millions of instructions per second.

MMS – Multimedia message service. Technology used in cellular telephony for the transmission of data, such as text, image, audio, and video.

Mobile broadband – Type of broadband connection. Wireless long distance access technology through a mobile phone network, especially 3G and 4G (respectively the third and fourth generation of mobile telephone standards defined by the International Telecommunication Union (ITU)).

Mobile code – Type of code used by web developers to incorporate greater functionality and improve the appearance of web pages. Some types of mobile code are: Java programs and applets, JavaScripts, and ActiveX components (or controls).

Mobile device – Equipment with computational resources that, due to its small size, offers great mobility of use and can be easily carried by its owner. Examples: notebooks, netbooks, tablets, PDAs, smartphones, and cell phones.

Modem – MOdulator and DEModulator. Equipment that converts digital signals from a computer into analog signals that can be transmitted by physical means, such as telephone lines, pay-TV cable, or fiber optics.

MPEG – Motion Picture Experts Group. Organizations which, among other things, define standards for video signals.

MSB – Most significant bit. Binary digit in the most significant position of the binary word.

MSE – Mean square error or average error.

Multiple domains (MDC) – Certificate for companies that have multiple addresses and want to encrypt all of them.

MUX – Multiplexer. Equipment that combines several signals to be transmitted on a single transmission channel.

Netiquette – Set of standards of conduct for Internet users, defined in the document "RFC 1855: Netiquette Guidelines."

Network scanning – Technique that consists of performing thorough searches on networks, with the objective of identifying active computers and collecting information about them, such as available services and installed programs.

NEXT – Near-end crosstalk. Crosstalk or interference between local transmitter and receiver, usually associated with a pair cable.

NTP – Network Time Protocol. Type of protocol that allows the synchronization of the clocks of devices on a network, such as servers, workstations, routers, and other equipment, from reliable time references (Source: http://ntp.br/).

NTT – Nippon Telephone and Telegraph Co.

OFDM – Orthogonal frequency division multiplexing.

Open Proxy – Poorly configured proxy that can be abused by attackers and used as a way to make certain actions on the Internet anonymous, such as attacking other networks or sending spam.

Opt-in – Message sending rule that defines that it is prohibited to send commercial emails or spam, unless there is a prior agreement by the recipient.

Opt-out – Rule for sending messages that defines that it is allowed to send commercial e-mails, but a mechanism must be provided so that the recipient can stop receiving the messages.

OSPF Open Shortest Path First. A type of dynamic routing protocol.

P2P – Acronym for peer-to-peer. Network architecture where each computer has equivalent functionalities and responsibilities. It differs from the client/server architecture, in which some devices are dedicated to serving others. This type of network is usually implemented via P2P programs, which allow [connecting one user's computer to another to share or transfer data, such as MP3, games, videos, and images.

Page defacement – Also known as graffiti, it is a technique that consists of altering the content of the web page of a website.

Password – Character set, unique to the user, used in the process of verifying his identity, ensuring that he is really who he claims to be and that he has the right to access the resource in question.

Patch – Security fix.

PCI – Standards established by global financial institutions to ensure security in credit card payment transactions.

PenTest – Test that simulates an invasion on the website in order to find and correct the identified security flaws.

Personal firewall – Device used to protect only one computer from unauthorized access.

Personal firewall – Specific type of firewall. Program used to protect a computer from unauthorized access from the Internet.

PGP – Pretty Good Privacy. Program that implements symmetric key cryptography, asymmetric keys, and digital signature. It has commercial and free versions.

Pharming – Specific type of phishing that involves redirecting the user's navigation to fake websites, through changes to the DNS service.

Phishing – Type of scam through which a scammer tries to obtain a user's personal and financial data, through the combined use of technical means and social engineering.

Plug-in – Program generally developed by third parties and which can be installed in the web browser or e-mail programs to provide extra functionality.

PMD – Principle of majority decision. When receiving a digital signal, the principle guarantees the choice of the most frequent symbol.

PN – Pseudorandom noise. Digital noise generated by the displacement recorder widely used in systems with spectral spreading by direct sequence.

POP3 Post Office Protocol version 3 is used by client users to access an e-mail account on a server to get the message.

Pop-up window – Type of window that appears automatically and without permission, overlapping the web browser window, after the user accesses a website.

PPP Point-to-Point Protocol is a form of serial data encapsulation that represents an improvement over SLIP, capable of providing bidirectional serial communication similar to SLIP but supports AppleTalk, IPX, TCP/IP, and NetBEUI along with TCP/IP.

Prepayment fraud – Type of fraud in which a scammer tries to trick a person into providing confidential information or making an advance payment, with the promise of receiving some kind of benefit in the future.

Proxy – Server that acts as an intermediary between a client and another server. It is usually used in companies to increase the performance of access to certain services or to allow more than one machine to connect to the Internet. When badly configured (open proxy), it can be abused by attackers and used to make certain actions on the Internet anonymously, such as attacking other networks or sending spam.

PSTN – Public switched telephone network (PSTN).

QoS – Quality of service. Parameter used to measure the quality of various functions of a communication network.

RACE – Research and Development for Advanced Communications. European committee for the creation of advanced communication networks.

RAM – Random access memory.

RARP Reverse Address Resolution. This protocol is to allow a permanent memory computer to have an IP address starting from its Ethernet address.

RENPAC – Switched network, X.25 standard, used in Brazil in the 1970s.

RIP Routing Information Protocol dynamically updates routing tables over WANs and the Internet. The distance vector algorithm is used to calculate the best route for the packet.

ROM – Read-only memory.

Rootkit – Type of malicious code. Set of programs and techniques that allows you to hide and ensure the presence of an attacker or other malicious code on a compromised computer. It is important to note that the name rootkit does not indicate that the tools that comprise it are used to obtain privileged access (root or Administrator) on a computer, but rather to maintain privileged access on a previously compromised computer.

RS – Reed-Solomon. One of the classes of error correction codes.

RZ – Return-to-zero. Coding system in which part of a symbol takes the zero level.

SAN – Small area network or small computer network, usually limited to a few tens of meters.

San UCC – Certified for Microsoft Exchange applications.

SA – Simulated annealing. Algorithm used in optimization problems. It relates optimization problems of a combinatorial nature with the physical process of cooling molten metals.

Scam – Misleading or fraudulent schemes or actions. Usually, they aim to obtain financial advantages.

Scanner – Program used to scan computer networks in order to identify which computers are active and which services are being made available by them. Widely used by attackers to identify potential targets, as it allows to associate possible vulnerabilities with services enabled on a computer.

Scan/Scanner – Technique for scanning networks and computers, in order to find viruses or unauthorized modifications.

Screenlogger – Specific type of spyware. Program similar to keylogger, capable of storing the position of the cursor and the screen displayed on the monitor, when the mouse is clicked, or the region surrounding the position where the mouse is clicked. It is widely used by attackers to capture the keys typed by users on virtual keyboards, available mainly on Internet banking sites.

Secure connection – Connection that uses an encryption protocol for data transmission, such as HTTPS or SSH.

Security fix – Fix designed to eliminate security holes in a program or operating system.

Security incident – Any adverse event, confirmed or suspected, related to the security of computer systems or computer networks.

Security policy – Document that defines the rights and responsibilities of each person in relation to the security of the computational resources he uses and the penalties to which he is subject to, in case he does not comply with it.

Site – Internet site identified by a domain name, consisting of one or more hypertext pages, which may contain text, graphics, and multimedia information.

Site shielding – Service package that identifies and corrects website vulnerabilities.

SLIP – Serial Line Internet Protocol – This protocol places data packets in frames in preparation for transport by network equipment. It sends data over serial lines, without error correction, addressing, or packet identification and without authentication. SLIP only supports the transport of IP packets.

S/MIME – Secure/multipurpose Internet mail extensions. Standard for signing and encrypting e-mails.

SMS – Short message service is a technology used in telephony to send and receive short text messages. Unlike MMS, it only allows text data and each message is limited to 160 alphanumeric characters.

SMTP – Simple Mail Transfer Protocol. Protocol responsible for transporting e-mail messages on the Internet.

Sniffer – Computer device or program used to capture and store data traveling over a computer network. It can be used by an attacker to capture sensitive information (such as user passwords), in cases where insecure connections are being used, that is, without encryption.

Sniffing – Traffic interception.

SNMP – Simple Network Management Protocol is used to manage network elements based on sending and receiving data.

SNRseg – Segmental signal-to-noise ratio.

SNR – Signal-to-noise ratio.

Social engineering – Technique by which one person seeks to persuade another to perform certain actions. Used by scammers to try to exploit greed, vanity, and good faith or to abuse the ingenuity and trust of others, in order to scam, deceive, or obtain confidential and important information.

Social network – Type of social network that allows users to create profiles and use them to connect with other users, share information, and group according to common interests. Some examples are: Facebook, Orkut, Twitter, Linkedin, Google+, and foursquare.

Soft opt-in – Rule similar to opt-in, but, in this case, it provides for an exception when there is already a business relationship between sender and recipient. Thus, the recipient's explicit permission is not required to receive e-mails from that sender.

SONET – Synchronous Optical NETworkNorth American standard for high-speed digital optical networks.

Spam bots – Automated programs that generate excessive traffic to a website through robots.

Spamcop – Institution that offers several anti-spam services, the most well-known being the one that allows you to automatically complain about incoming spam.

Spammer – Individual who sends the spam.

Spam – Term used to refer to unsolicited e-mails, which are usually sent to a large number of people.

Spam zombie – Computer infected by malicious code (bot), capable of transforming the user's system into an e-mail server for sending spam. In many cases, the user of the infected computer takes time to realize that his computer is being used by an attacker for this purpose.

Sponsored link – Type of link displayed prominently on a search site when specific words are searched by the user. When the user clicks on a sponsored link, the search engine receives a previously agreed value from the advertiser.

Spyware – Specific type of malicious code. Program designed to monitor the activities of a system and send the collected information to third parties. Keylogger, screenlogger, and adware are some specific types of spyware.

SQNR – Signal-to-quantization noise ratio.

SSH – Secure Shell is a protocol for accessing a remote computer using encryption, which allows actions such as executing commands and transferring files.

SSID – Service set identifier is a unique set of characters that identifies a wireless network. The SSID differentiates one wireless network from another, and a client can normally connect to a wireless network only if it can provide the correct SSID.

SSL EV – Extended validation cryptography, which in addition to protecting information trafficked on the website, validates the company's corporate name and CNPJ.

SSL – Secure Sockets Layer. Like TLS, it is a protocol that, through encryption, provides confidentiality and integrity in communications between a client and a server and can also be used to provide authentication. SSL is a basic encryption for any e-commerce that aims to protect information trafficked on the website. It was developed by the Netscape team in the 1990s and later replaced by TLS.

SWIFT – Society for Worldwide Interbank Financial Telecommunications. Network developed to support the exchange of information from banks and financial institutions.

TCP/IP – Transmission Control Protocol/Internet Protocol. This term is widely used in Internet protocols.

TCP – Reliable connection used to control the management of applications at the service level between computers. It does both the sequential transport of data and the verification of its integrity.

Telnet – Command used to open a remote session on another computer. It is based on TCP for transport and is defined by RFC854.

TFTP Trivial File Transfer Protocol – Allows the transfer of files between two computers without the need for a login and password. More limited sense is little used.

TIA – Telecommunications Industry Association.

TLS – Transport Layer Security. Like SSL, it is a protocol that, through encryption, provides confidentiality and integrity in communications between a client and a server and can also be used to provide authentication.

Traffic interception – Technique that consists of inspecting data trafficked on computer networks, through the use of specific programs called sniffers.

Trojan horse – Type of malicious code. Program normally received as a gift (for example, virtual card, photo album, screen saver, and game) that, in addition to performing the functions for which it was apparently designed, also performs other functions, usually malicious and without the user's knowledge.

UBE – Unsolicited bulk e-mail. Term used to refer to unsolicited emails sent in large quantities.

UCE – Unsolicited commercial e-mail. Term used to refer to unsolicited commercial e-mails.

UDP – Unreliable connection used to control the management of applications at the service level between computers.

UHF – Ultra-high frequency. Frequency range from 300 MHz to 3 GHz.

UNESCO – United Nations Economic, Scientific and Cultural Organization.

URL – Universal resource locator. String of characters that indicates the location of a resource on the Internet.

User account – Also called "username" and "login name." Corresponds to the unique identification of a user on a computer or service.

Virus – Program or part of a computer program, usually malicious, that spreads by inserting copies of itself, becoming part of other programs and files. The virus depends on the execution of the host program or file so that it can become active and continue the infection process.

VPN – Virtual private network. Term used to refer to the construction of a private network using public networks (for example, the Internet) as infrastructure. These systems use encryption and other security mechanisms to ensure that only authorized users can access the private network and that no data will be intercepted while passing through the public network.

Vulnerability Exploitation – A program or part of a malicious program designed to exploit an existing vulnerability in a computer program.

Vulnerability – Security holes that can be exploited by attackers.

WAF – Security firewall that protects web applications from attacks and invasions.

WAN – Wide area network. Computer network covering a wide area.

WAP – Wireless Application Protocol. Global protocol used on many wireless systems, for users to view and interact with data services.

Web bug – Image, usually very small and invisible, that is part of a web page or an e-mail message and that is designed to monitor who is accessing that page or e-mail message.

WEP – Wired equivalent privacy. Security protocol for wireless networks which implements encryption for data transmission.

Whitelist – List of e-mails, domains, or IP addresses, previously approved and that, normally, are not submitted to the configured antispam filters.

Wi-Fi – Wireless fidelity. Trademark generically used to refer to wireless networks that use any of the 802.11 standards.

Wildcard – Encryption for those who want to protect several subdomains under a specific address.

Wireless network – Network that allows the connection between computers and other devices through the transmission and reception of radio signals.

Wireless network – Wireless network.

WLAN – Wireless local area network. Type of network that uses high-frequency radio waves, instead of cables, for communication between computers.

WLL – Wireless local loop or wireless LAN. Broadband system that uses RF to transmit voice and data.

Worm – Type of malicious code. Program capable of automatically propagating through networks, sending copies of itself from computer to computer. Unlike the virus, the worm does not include its own copies in other programs or files and does not need to be run to spread. Its spread occurs through the exploitation of existing vulnerabilities or flaws in the configuration of programs installed on computers.

WPA – Wi-Fi Protected Access. Security protocol for wireless networks designed to replace the WEP protocol. Designed to operate with Wi-Fi products that only offer WEP technology, through program updates. It includes two improvements over WEP: better encryption for data transmission and user authentication.

X.25 – Standard of data transmission in packets.

Zombie computer – Name given to a computer infected by bot, as it can be controlled remotely, without the knowledge of its owner.

Zombie-computer – Zombie computer, infected by a bot.

References

Abramson, N. (1963). *Information Theory and Coding*. McGraw-Hill Book Co., New York.

Abramson, N. (1963). *Information Theory and Coding*. McGraw-Hill, New York.

Aczél, J. and Daróczy, Z. (1975). *On Measures of Information and their Characterizations*. Academic Press, New York.

Adke, S. R. and Manjunath, S. M. (1984). *An Introduction to Finite Markov Processes*. John Wiley and Sons, New Delhi.

Akkouchi, M. (2008). On the Convolution of Exponential Distributions. *Journal of the Chungcheong Mathematical Society*, 21(4):501–510.

Alencar, M. S. (1999). *Princípios de Comunicações*. Editora Universitária, UFPB, João Pessoa, Brasil.

Alencar, M. S. (2001). *Sistemas de Comunicações*. Editora Érica Ltda., ISBN 85-7194-838-0, São Paulo, Brasil.

Alencar, M. S. (2002). *Telefonia Digital, Quarta Edição*. Editora Érica Ltda., ISBN 978-85-365-0364-6, São Paulo, Brasil.

Alencar, M. S. (2007). The Information is Relative – Second Part (in Portuguese). Article for an electronic journal in the Internet, Jornal do Commercio *On Line*, Recife, Brazil.

Alencar, M. S. (2008a). Como era o mundo sem internet – I. Artigo para jornal eletrônico na Internet, Jornal do Commercio *On Line*, Recife, Brasil.

Alencar, M. S. (2008b). Como era o mundo sem internet – II. Artigo para jornal eletrônico na Internet, Jornal do Commercio *On Line*, Recife, Brasil.

Alencar, M. S. (2008c). Como era o mundo sem internet – III. Artigo para jornal eletrônico na Internet, Jornal do Commercio *On Line*, Recife, Brasil.

Alencar, M. S. (2008d). Como era o mundo sem internet – IV. Artigo para jornal eletrônico na Internet, Jornal do Commercio *On Line*, Recife, Brasil.

Alencar, M. S. (2010). *Divulgação Científica*. Gráfica e Editora Epgraf, ISBN 978-85-910418-2-4, Campina Grande, Brasil.

Alencar, M. S. (2011a). Evolução da Internet. *Revista de Tecnologia da Informação e Comunicação*, 1(1):6–10.

Alencar, M. S. (2011b). *História da Comunicação no Brasil*. Gráfica e Editora Epgraf, ISBN 978-85-910418-3-1, Campina Grande, Brasil.

Alencar, M. S. (2011c). *História, Tecnologia e Legislação de Telecomunicações*. Gráfica e Editora Epgraf, ISBN 978-85-910418-4-8, Campina Grande, Brasil.

Alencar, M. S. (2012a). *Engenharia de Redes de Computadores*. Editora Érica Ltda., ISBN 978-85-365-0411-7, São Paulo, Brasil.

Alencar, M. S. (2012b). *Engineering of Computer Networks (in Portuguese)*. Publisher Editora Érica Ltda., ISBN 978-85-365-0411-7, São Paulo, Brazil.

Alencar, M. S. (2014a). Espionagem entre Parceiros. Artigo para jornal eletrônico na Internet, NE10 – Sistema Jornal do Commercio de Comunicação, Recife, Brasil.

Alencar, M. S. (2014b). *Teoria de Conjuntos, Medida e Probabilidade*. Editora Érica Ltda., ISBN 978-85-365-0715-6, São Paulo, Brasil.

Alencar, M. S. (2015). *Informação, Codificação e Segurança de Redes*. Editora Elsevier, ISBN: 978-85-352-8184-2, Rio de Janeiro, Brasil.

Alencar, M. S. (2017a). A Internet das Coisas. Artigo para jornal eletrônico na Internet, NE10 – Sistema Jornal do Commercio de Comunicação, Recife, Brasil.

Alencar, M. S. (2017b). As Coisas da Internet. Artigo para jornal eletrônico na Internet, NE10 – Sistema Jornal do Commercio de Comunicação, Recife, Brasil.

Alencar, M. S. (2017c). Big Data. Artigo para jornal eletrônico na Internet, NE10 – Sistema Jornal do Commercio de Comunicação, Recife, Brasil.

Alencar, M. S. (2017d). Como ser Reconhecido. Artigo para jornal eletrônico na Internet, NE10 – Sistema Jornal do Commercio de Comunicação, Recife, Brasil.

Alencar, M. S. (2017e). O Reconhecimento na Ponta dos Dedos. Artigo para jornal eletrônico na Internet, NE10 – Sistema Jornal do Commercio de Comunicação, Recife, Brasil.

Alencar, M. S. (2018a). A Urna, o Código e a Ética. Artigo para jornal eletrônico na Internet, NE10 – Sistema Jornal do Commercio de Comunicação, Recife, Brasil.

Alencar, M. S. (2018b). Criptografia para Iniciantes. Artigo para jornal eletrônico na Internet, NE10 – Sistema Jornal do Commercio de Comunicação, Recife, Brasil.

Alencar, M. S. (2018c). Segurança de Redes. Artigo para jornal eletrônico na Internet, NE10 – Sistema Jornal do Commercio de Comunicação, Recife, Brasil.

Alencar, M. S. (2021). *Economic Science (print)*. River Publishers, Delft, The Netherlands.

Alencar, M. S. and Alencar, R. T. (2016). *Probability Theory*. Momentum Press, LLC, ISBN-13: 978-1-60650-747-6 (print), New York, USA.

Alencar, M. S. and Assis, K. D. R. (2020). A New Derivation of the Leftover Hash Lemma. *The ISC International Journal of Information Security*, 12(3):55–58.

Ash, R. B. (1965). *Information Theory*. Dover Publications, Inc., New York, USA.

Baldwin, C. (2016). *Bitcoin Worth US$72 Million Stolen from Bitfinex Exchange in Hong Kong*. Reuters, New York, USA.

Bandeira, L. A. M. (2007). *Presença dos Estados Unidos no Brasil*. Civilização Brasileira, São Paulo, Brasil.

Berens, S. (2013). Conditional Rényi Entropy. Master's thesis, Mathematisch Institute, Universiteit Leiden.

Blake, I. F. (1987). *An Introduction to Applied Probability*. Robert E. Krieger Publishing Co., Malabar, FL, USA.

Boyer, C. (1974). *História da Matemática*. Editora Edgard Blucher Ltda., São Paulo, Brasil.

Camacho, T. S. and da Silva, G. J. C. (2018). Criptoativos: Uma AnÃ¡lise do Comportamento e da Formação do Preço do Bitcoin. *Revista de Economia*, 39(68):1–26.

Carter, J. L. and Wegman, M. N. (1979). Universal Classes of Hash Functions. *Journal of Computer and System Sciences*, 18:143–154.

CERT.br. (2019). Glossário. Cartilha de Segurança para Internet, Centro de Estudos, Resposta e Tratamento de Incidentes de Segurança no Brasil, São Paulo, Brasil.

Csiszár, I. and Kórner, J. (1981). *Information Theory: Coding Theorems for Discrete Memoryless Systems*. Academic Press, New York, USA.

Communications System Design (2001). Communication systems design – reference library: Glossary. Internet site, www.csdmag.com/glossary.

Dudeney, H. E. (1924). Puzzle: Send More Money. *Strand Magazine*, (68):97 and 214.

Farouzan, B. A. (2008). *Comunicação de Dados e Redes de Computadores, Quarta Edição.* AMGH Editora Ltda., São Paulo, Brasil.

Feinstein, A. (1958). *Foundations of Information Theory.* McGraw-Hill Book Company, Inc., New York, USA.

Forouzan, B. and Mosharraf, F. (2011). *Computer Networks: A Top Down Approach.* McGraw-Hill, New York, USA.

Furiati, G. (1998). Serviços são Reclassificados. *Revista Nacional de Telecomunicações,* 227:32–35.

Gast, M. S. (2005). *802.11 Wireless Networks – The Definitive Guide.* O'Reilly, USA.

Giozza, W. F., Araújo, J. F. M, ao de Moura, J. A., and Sauvé, J. P. (1986). *Redes Locais de Computadores – Tecnologia e Aplicações.* McGraw-Hill, Ltda., Rio de Janeiro, Brasil.

Halmos, P. R. (1960). *Naive Set Theory.* D. Van Nostrand Company, Inc., Princeton, USA.

Hammond, J. L. and O'Reilly, P. J. P. (1986). *Performance Analysis of Local Computer Networks.* Addison-Wesley Publishing Company, Reading, USA.

Hartley, R. V. L. (1928). Transmission of Information. *Bell Systems Technical Journal,* 535.

Hayes, J. F. (1986). *Modeling and Analysis of Computer Communications Networks.* Plenum Press, New York, USA.

Haykin, S. (1988). *Digital Communications.* John Wiley and Sons, New York, USA.

Haykin, S. (1999). *The Code Book – The Science of Secrecy from Ancient Egypt to Quantum Cryptography.* Doubleday, New York, USA.

Hosch, W. L., Jain, P., and Rodriguez, E., editors (2021). *Cryptography.* Encyclopaedia Britannica, Inc., Chicago, USA.

IBM, editor. (2021). *Cryptography.* IBM Corporation, New York, USA.

IEEE. (2004). *IEEE Standard for Local and Metropolitan Area Networks Part 16: Air Interface for Fixed Broadband Wireless Access Systems, IEEE Standard, 801.16.* The Institute of Electrical and Electronics Engineers, Inc., United States of America.

Global Initiative (2017). 10 biggest cyber crimes and data breaches to date. [Online; Access on June 30, 2021].

James, B. R. (1981). *Probabilidade: Um Curso em Nível Intermediário.* Instituto de Matemática Pura e Aplicada – CNPq, Rio de Janeiro, Brasil.

Joe Tidy. (2021). Pegasus: Spyware sold to governments 'targets activists'. [Online; Access on July 23, 2021].

Kaspersky Laboratory (2021). Tips on how to protect yourself against cybercrime. [Online; Access on June 30, 2021].

Kendall, D. G. (1953). Stochastic Processes Occurring in the Theory of Queues and their Analysis by the Method of the Imbedded Markov Chain. *The Annals of Mathematical Statistics*, 24(3):338–354.

Kessler, G. C. (2021). *An Overview of Cryptography*. Daytona Beach, FL, USA. Available at: https://www.garykessler.net/library/crypto.html.

Khinchin, A. I. (1957). *Mathematical Foundations of Information Theory*. Dover Publications, Inc., New York, USA.

Kleinrock, L. (1975). *Queuing Systems*. John Wiley & Sons, New York, USA.

Leon-Garcia, A. (1989). *Probability and Random Processes for Electrical Engineering*. Addison-Wesley Publishing Co., Reading, MA. USA.

Leon-Garcia, A. and Widjaja, I. (2000). *Communication Networks*. McGraw-Hill, Boston, MMA, USA.

Lima, V. R. D., Lijó, M. C., Sousa, M. P. (2014). Segurança em Redes sem Fio: PrincÃpios, Ataques e Defesas. *Revista de Tecnologia da Informação e Comunicação*, 4(2):1–10.

Lipschutz, S. (1968). *Teoria de Conjuntos*. Ao Livro Técnico S.A., Rio de Janeiro, Brasil.

Little, J. D. C. (1961). A Proof for the Queuing Formula: $L = \lambda W$. *Operations Research*, 9(3):383–387.

MacKay, D. J. C. (2003). *Information Theory, Inference, and Learning Algorithms*. Cambridge University Press, Cambridge, U.K.

Mangion, D. (2019). The 10 worst cyber crimes analysed. [Online; Access on June 30, 2021].

Alencar, M. S., Carvalho, F. B. S., Lopes, W. T. A., and Markarian, G. (2006). Model for the Transmission Rate versus File Length for a PLC System. In *The Second International Symposium on Broadband Communications (ISBC'06)*, pp. 81–83, Moscow, Russia.

Markovic, M. (2002). "Data Protection Technics and Cryptographic Protocols in Modern Computer Networks". Course notes, Mathematical Institute SANU, Beograd, Serbia and Montenegro.

Marksteiner, S., Jiménez, V. J. E., Vallant, H., and Zeiner, H. (2017). An Overview of Wireless IoT Protocol Security in the Smart Home Domain. In *Proceedings of the 2017 Internet of things Business Models, Users, and Networks*, pp. 1–9.

Massey, J. L. (1990). The Relevance of Information Theory to Modern Cryptography. In *Proceedings of the Bilkent International Conference on New Trends in Communications, Control and Signal Processing (NILCON'90)*, pp. 176–182, Elsevier Science Publisher, Ankara, Turkey.

Maurer, U. M. (1989). A Provably-Secure Strongly-Randomized Cipher. In *Proceedings of the Monte Verita Seminar On Future Directions in Cryptography*, Ascona, Switzerland.

Maurer, U. M. (1995). Generalized Privacy Amplification. *IEEE Transactions on Information Theory*, 41(6):122–127.

Mediacenter, P. (2021). Types of cybercrime. [Online; Access on June 30, 2021].

MobileWord. (2001). Mobileword's glossary. Internet site, www.mobileworld.org.

Moura, J. A., Sauvé, J. P., Giozza, W. F., and Araújo, J. F. M. (1986). *Redes Locais de Computadores – Protocolos de Alto Nível*. McGraw-Hill, Ltda., Rio de Janeiro, Brasil.

Nyquist, H. (1924). Certain Facts Affecting Telegraph Speed. *Bell Systems Technical Journal*, pp. 324.

Papoulis, A. (1983). Random Modulation: A Review. *IEEE Transactions on Accoustics, Speech and Signal Processing*, 31(1):96–105.

Pierce, John R. (1980). *An Introduction to Information Theory – Symbols, Signals & Noise, Second Edition*. Dover Publications, Inc., New York, USA.

Pouw, K. D. (1999). "Segurança na Arquitetura TCP/IP: de *Firewalls* a Canais Seguros". Dissertação de mestrado, Departamento de Engenharia Elétrica, Universidade Estadual de Campinas, Campina, Brasil.

Preneel, B. (1994). Cryptographic Hash Functions. *European Transactions on Telecommunications*, 5(4):431–448.

Queiroz, D.V., Vieira, J. C. M., and Fonseca, I. E., (2014). Deteção de Ataques de Negação de Serviço Utilizando Ferramentas de Monitoramento e Análise de Tráfego. *Revista de Tecnologia da Informação e Comunicação*, 4(1):1–8.

Rényi, A. (1961). On Measures of Entropy and Information. In *Proceedings of the 4th Berkeley Symposium on Mathematics, Statistics and Probability*, Vol. 1, pp. 547–561, University of California Press, Berkeley, USA.

Reza, F. M. (1961). *An Introduction to Information Theory*. McGraw-Hill Book Co., New York, USA.

Ricciardi, L. M. (1990). *Lectures in Applied Mathematics and Informatics*. Manchester University Press, Manchester, U.K.

Rocha Jr., V.C. da (2013). Segurança de Rede. *Revista de Tecnologia da Informação e Comunicação*, 3(1):14–21.

Sayood, K. (2006). *Introduction to Data Compression*. Morgan Kaufmann, San Francisco.

Shannon, C. E. (1948a). A Mathematical Theory of Communication. *The Bell System Technical Journal*, 27:379–423.

Shannon, C. E. (1948b). A Mathematical Theory of Communication. *The Bell System Technical Journal*, 27:379–423.

Shannon, C. E. (1949). Communication Theory of Secrecy Systems. *The Bell System Technical Journal*, 4(28):656–715.

Skycell. (2001). Glossary of satellite terminology. Internet site, www.satellit etelephone.com.

Soares, L. F. G., Lemos, G., and Colaher, S. (1995). *Redes de Computadores*. Editora Campus, Rio de Janeiro, Brasil.

Stallings, W. (1999). *Cryptography and Network Security – Principles and Practice*. Prentice Hall, Upper Saddle River, USA.

Stinson, D. R. (1994a). Combinatorial Techniques for Universal Hashing. *Journal of Computer and System Sciences*, 48:337–346.

Stinson, D. R. (1994b). Universal Hashing and Authentication Codes. *Designs, Codes and Cryptography*, 4:369–380.

Tanenbaum, A. S. (1989). *Computer Networks*. Prentice-Hall, Englewood Cliffs, USA.

Tanenbaum, A. S. (2003). *Computer Networks*. Prentice-Hall, PTR, Englewood Cliffs, USA.

TIAB2B.com. (2001). Everything communications. Internet site, www.tiab 2b.com/glossary.

van der Lubbe, J. C. A. (1997). *Information Theory*. Cambridge University Press, Cambridge, U.K.

Whittle, P. (2007). *Networks: Optimization and Evolution*. Cambridge University Press, Cambridge, UK.

Wikipedia contributors. (2019). Cryptographic hash function — Wikipedia, the free encyclopedia. [Online; Access on 16 de julho de 2019].

Wikipedia contributors (2021). Pegasus (spyware) — Wikipedia, the free encyclopedia. [Online; Access on July 23, 2021].

Wilde, J., Merritt, M., Freyman, N., Qin, S., and Yu, Y. (2021a). The biggest ransomware attack ever. [Online; Access on July 6, 2021].

Wilde, J., Merritt, M., Freyman, N., Qin, S., and Yu, Y. (2021b). The top five business stories of q2. [Online; Access on June 30, 2021].

Zadeh, L. A. (1965). Fuzzy Sets. *Information and Control*, 8(3):338–353.

Zhao, W. (2019). *Crypto Exchange Bithumb Hacked for US$ 13 Million in Suspected Insider Job*. CoinDesk, New York, USA.

Index

About the Author

Marcelo Sampaio de Alencar.

Marcelo Sampaio de Alencar was born in Serrita, Brazil, in 1957. He received the bachelor's degree in electrical engineering from the Federal University of Pernambuco (UFPE), Brazil, in 1980, the master's degree in electrical engineering, from the Federal University of Paraiba (UFPB), Brazil, in 1988, and the Ph.D. degree from the University of Waterloo, Department of Electrical and Computer Engineering, Canada, in 1993. He has more than 40 years of engineering experience and more than 30 years as an IEEE Member. He is currently a Senior Member. Between 1982 and 1984, he worked with the State University of Santa Catarina (UDESC). From 1984 to 2003, he worked with the Department of Electrical Engineering, Federal University of Paraiba, where he was a Full Professor and supervised more than 70 graduate and several undergraduate students. From 2003 to 2017, he was a Chair Professor with the Department of Electrical Engineering, Federal

University of Campina Grande, Brazil. He also spent some time working, as a consultant, for MCI-Embratel, and for the University of Toronto, as Visiting Professor. He was a Visiting Chair Professor with the Department of Electrical Engineering, Federal University of Bahia, from 2017 to 2019. He was a Visiting Researcher with the Senai Cimatec, Salvador, Bahia, Brasil, from 2019 to 2021.

He is the founder and president of the Institute for Advanced Studies in Communications (Iecom). He has been awarded several scholarships and grants, including three scholarships and several research grants from the Brazilian National Council for Scientific and Technological Research (CNPq), two grants from the IEEE Foundation, a scholarship from the University of Waterloo, a scholarship from the Federal University of Paraiba, an achievement award for contributions to the Brazilian Telecommunications Society (SBrT), an academic award from the Medicine College of the Federal University of Campina Grande (UFCG), and an achievement award from the College of Engineering of the Federal University of Pernambuco, during its 110th year celebration. He is a laureate of the 2014 Attilio Giarola Medal.

He published over 500 engineering and scientific papers and 28 books: *Cryptography and Network Security, Economic Theory, Music Science, Linear Electronics, Modulation Theory, Scientific Style in English,* and *Cellular Network Planning* by River Publishers, *Spectrum Sensing Techniques and Applications, Information Theory, and Probability Theory* by Momentum Press, *Information, Coding and Network Security* (in Portuguese) by Elsevier, *Digital Television Systems* by Cambridge, *Communication Systems* by Springer, *Principles of Communications* (in Portuguese) by Editora Universitária da UFPB, *Set Theory, Measure and Probability, Computer Networks Engineering, Electromagnetic Waves and Antenna Theory, Probability and Stochastic Processess, Digital Cellular Telephony, Digital Telephony, Digital Television and Communication Systems* (in Portuguese) by Editora Érica Ltda, and *History of Communications in Brazil, History, Technology and Legislation of Communications, Connected Sex, Scientific Diffusion, Soul Hiccups* (in Portuguese) by Epgraf Gráfica e Editora. He also wrote several chapters for 11 books. His biography is included in the following publications: *Who's Who in the World* and *Who's Who in Science and Engineering,* by Marquis Who's Who, New Providence, NJ, USA.

Marcelo S. Alencar has contributed in different capacities to the following scientific journals: Editor of the *Journal of the Brazilian*

Telecommunication Society; Member of the International Editorial Board of the *Journal of Communications Software and Systems* (JCOMSS), published by the Croatian Communication and Information Society (CCIS); Member of the Editorial Board of the *Journal of Networks* (JNW), published by Academy Publisher; Founder and Editor-in-Chief of the *Journal of Communication and Information Systems* (JCIS), special joint edition of the IEEE Communications Society (ComSoc) and SBrT. He is a member of the SBrT-Brasport Editorial Board. He has been involved as a volunteer with several IEEE and SBrT activities, including being a member of the Advisory or Technical Program Committee in several events. He served as a member of the IEEE Communications Society Sister Society Board and as liaison to Latin America Societies. He also served on the Board of Directors of IEEE's Sister Society SBrT. He is a Registered Professional Engineer. He was a columnist of the traditional Brazilian newspaper Jornal do Commercio, from 2000 to 2019, and he was vice-president external relations of SBrT. He is a member of the IEEE, IEICE, in Japan, and SBrT, SBMO, SBPC, ABJC, and SBEB, in Brazil. He studied acoustic guitar at the Federal University of Paraiba and keyboard and bass at the music school Musidom. He is a composer and percussionist of the carnival club *Bola de Ferro*, in Recife, Brazil.